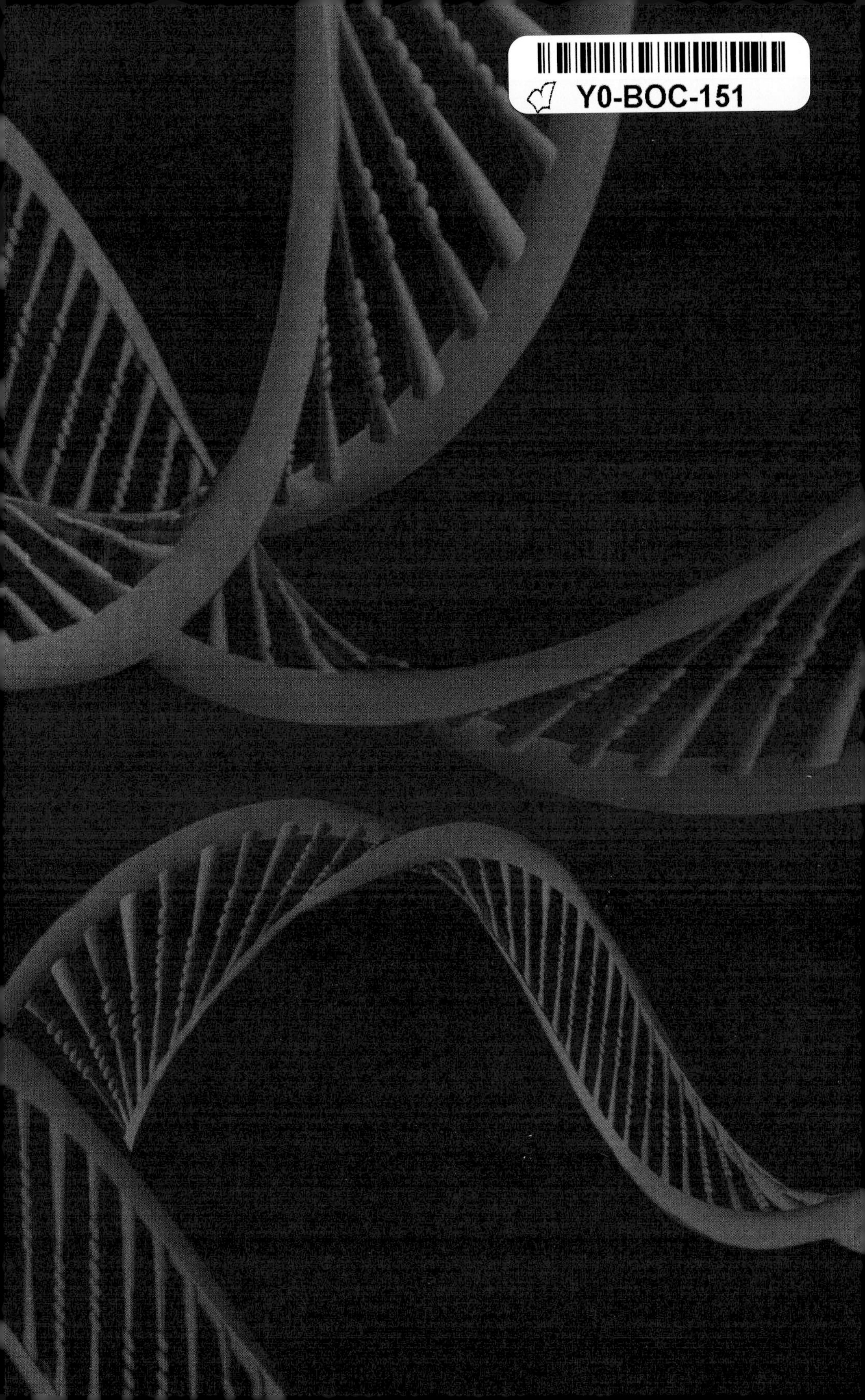
Y0-BOC-151

Beneath This Gruff Exterior

There Beats a Heart of Plastic

Table of Contents

Acknowledgments

My biographical anecdotes have been converted into readable form by the only person in the world who could have done it, my daughter Kimm. As a professional writer and humorist, she has been able to assemble my ramblings into some semblance of order.

Particularly in the chapter relating to business intrigue in biotech, I have changed names to protect the innocent (and the guilty). In much of the rest of the book the names are real. Regardless of whether the names are actual or imagined, the events are all as accurate as I remember them.

Acknowledgments

Vitals

You are about to read a classic rags-to-riches story. Alan Walton, born to a humble English family, grew up to enjoy not just material wealth but also riches of the spirit, in the form of significant accomplishments in the life sciences sprinkled with dashes of adventure.

Here are a few facts to put that story into context.

Walton was born in 1936 in Kings Norton, a small village just outside of Birmingham, England. He went to Kings Norton Boys' Grammar School, and then Nottingham University, where he majored in chemistry. He got his Ph.D. when he was 23.

He came to the United States in 1960 and worked at Indiana University for a little while. Then he moved on to Case Western Reserve and worked there for several years. He was the youngest person at Case ever to be appointed a full tenured professor. He published 150 articles and books, and wound up as Director of the Laboratory for Biological Macromolecules. While he was at Case Western, he was invited to lecture all over the world, and he was a science advisor to President Jimmy Carter.

One of the projects he directed at Case involved artificial hearts and arteries, and his graduate students posted a picture of him on his door, with the caption "Beneath This Gruff Exterior There Beats a Heart of Plastic." Hence the name of this book.

In 1981 he left academia and founded one of the first biotech companies, University Genetics. By the time he left, UGEN was one of the handful of largest biotech companies in the world.

In 1987 Walton turned to venture capital, joining Oxford Partners. Since then he has raised and managed three venture capital funds dedicated to the life sciences as chairman of Oxford Bioscience Partners. Oxford is one of the largest, if not *the* largest, venture funds dedicated to life sciences. That's what he's doing now.

Walton married Jasmin in 1959, and they had two children, Kimm and Keir. Jasmin died in 1970. For the last 22 years he has been married to E.J., who brought to the family two daughters of her own, Kristin and Sherri.

— The Publishers

Foreword

What a stupid way to die. What a stupid, stupid, stupid way to die!

A few months ago, I was fishing from a canoe in the small lake behind a friend's house. Over the years I've dabbled in skydiving, flying acrobatics, mountain climbing—activities whose inherent riskiness is much of their attraction, along with the adrenaline rush and the ability to brag about them afterwards. But fishing—no, there's nothing life-threatening about fishing. And that particular day my fishing trip had no hallmarks of danger. I was in a canoe in water no more than ten feet deep, and only thirty feet from the shore.

But then a wind gust caught me off-balance and dumped me into the water. My windbreaker ballooned behind me and pinned my arms, effectively handcuffing me a few pathetic feet from safety. My first thought was: this is no way to die!

But I couldn't free my arms. I was weighted down by wet clothes. The water was very cold. I tried treading water, but I quickly tired. My face bobbed beneath the surface and I felt myself losing consciousness. I was slipping under for what I believed would be the last time.

It was only the most gossamer web of coincidences that saved my life.

Henry, my daughter Kimm's boyfriend, had providentially decided to come fishing with me that afternoon. He was sitting on a dock some distance away. He couldn't see me floundering in the water, but a woman in a backyard nearby noticed me. She yelled over to Henry, "Hey! I think your friend is in trouble!" He ran over to help.

He picked up a long stick and held it out toward me, telling me to grab on to it. I didn't hear him; I didn't respond. As a certified lifeguard, he knew exactly what to do. He dived in, fully-clothed, and swam up behind me. Just as I expected to be swallowed up by the lake, I felt him lift me up from behind and drag me toward the shore.

I am delighted to report that he is now my son-in-law.

That "near death" experience spurred me to do something I'd long wanted to do. That is, pen a memoir. More specifically, to try to emulate the book *Surely You're Joking, Mr. Feynmann,* by the hero of my younger life, the Nobel Laureate and Cal Tech physics professor Richard Feynmann.

Feynmann recounted his life in that book by sharing anecdotes about his adventures, observations about the world around him, and reflections on the work he did. Feynmann didn't tell many stories about his family. Similarly, my book does not describe the most important people in my life—my family—in great detail. Just as Feynmann did, I've put together an informal collection of stories placed in the vague stream of history.

I can't promise my own recollections will be as sparkling as Feynmann's. But I was fortunate to be born into exciting times, find challenging and groundbreaking work, meet fascinating people, and marry a wonderful woman (after a couple of false starts).

This is my story.

Chapter One

Luck of the War

Bananas.

In England, during the Second World War, fresh fruit was unheard of. Rationing was very stringent. We got four ounces of meat per week, very little sugar, and one egg (though later in the war we kept a few chickens that provided the luxury of more eggs). My meals at home mainly consisted of potatoes, gravy and boiled cabbage. At school the dinner was equally horrid. God knows what it was made from. Probably some permutation of potatoes, gravy and boiled cabbage. Whatever it was, it smelled and tasted just awful.

And never, whether at home or at school, did we get fresh fruit.

As I was only three when the war started, I had no recollection of a time when anything else was available.

Like bananas.

I remember hearing my parents and their friends rapturously describe bananas, and I looked forward to the day when I would actually get to taste one myself.

One day soon after the war my father came home with a paper-wrapped package, and, smiling, said to me, "Look what I've found for you, Alan. A *banana!*"

I watched with feverish anticipation as he unwrapped the package. There, in what looked like a specimen jar, was something that might once have resembled a banana, but didn't any longer. At the time, of course, I didn't know any better. All I saw was something that looked remarkably like a stool sample bobbing in a colourless fluid.

My father prodded me, and said, "It's for *you.*"

Not wanting to disappoint him—but also quite certain that I wasn't going to actually *eat* this thing—I called his bluff. "Oh, Dad, I wouldn't want to eat it all myself. Why don't I share it with you and Mom?"

So the three of us shared this hideous fruity slug. Its looks didn't belie its taste. It was something like eating a wet, tasteless sponge. My parents, lifelong practitioners of the British stiff upper lip, didn't betray what they thought of it.

It wasn't until 1948 that bananas started being shipped to England from the colonies, particularly Jamaica. The first time I saw one was when somebody pointed it out to me at a greengrocer's. "Ooh, look at the lovely bananas!"

"That's not a banana," I snorted. "Bananas aren't yellow. They're brown. And they're bloody awful."

xxxxxx

My memories of the war hardly involve the trauma of bloodshed. The closest I got to actually seeing death was when a horse was killed by a bomb in a nearby field. Once when I was walking home from school a German Messerschmidt fighter machine-gunned the street I was walking on. I jumped into the bushes. While the street was torn up rat-tat-tat, I wasn't hurt. And although a number of older houses burned to the ground within a quarter mile of our house, there were no fatalities.

Instead, the overwhelming feeling for me was a kind of second-hand fear, relayed by my parents. They had lived through World War I and had vivid memories of the gas attacks on the troops in the trenches. They often told me nightmare-inducing stories about people being disabled and killed by mustard gas. Our greatest fear came from our firm belief that the Germans would drop 'gas bombs' on us, something today we'd call chemical warfare. As a three- or four-year-old I had only the vaguest notion of what a 'gas bomb' was. I knew we all received gas masks—mine had Mickey Mouse on it—and we had to test them in a room that we were told contained gas, although I suspect now it really didn't.

I remember in the earliest days of the war that the German bombers would come over at night, and my mother and I would crouch under the dining room table, which she considered the safest place in our house. The sound of the German twin-engine Dornier and Heinkel bombers was so distinctive that it remains with me to this day: RRrrRRRrrrrRRRrrRRR, a kind of whirring rhythm of slightly out-of-phase engines.

We lived in Kings Norton, a small village near Birmingham. Kings Norton itself was of little interest to the Germans. When they flew over us, they had overflown Coventry, a town about twenty miles away where most of Britain's war aircraft were manufactured in pre-war automobile assembly plants. However, navigation aids weren't what they are now, and since the government enforced a total blackout—all windows had to be blacked out at night and all exterior lights extinguished—the Germans couldn't see exactly what they were aiming at. So they occasionally bombed our neighborhood, presumably by mistake.

As the bombings became more frequent and intense, the government provided us with what they called air raid shelters. These were actually pieces of curved corrugated steel that we were supposed to bolt together into a kind of outdoor metal tent. My father quickly decided this "shelter" wasn't much safer than our house. Instead, he dug a pit in our back yard, lined it with concrete, and formed a sturdier underground shelter. I helped him construct it, and we buried it under a rock garden. What should have been quite cozy was in practice dank, dark and somewhat terrifying. I remember nights, hiding in the shelter with my mother, as we heard bombs whistle and explode.

Fairly early on the Germans switched from high explosive bombs to "incendiary" bombs, which were designed to penetrate the roofs of houses and burn them down. These bombs were about nine inches long, which I know because one landed in my friend David Jackson's apple tree, lodging between two branches. The two of us were greatly excited by this and clambered up to take a closer look at it. In our childish innocence we didn't realize that it could have gone off while we were examining it. Fortunately for us, it didn't.

Predictably, the war impacted our life at school. There were savings bond drives, where we would take in a sixpence and get a stamp. And there were air raid shelters on the school grounds. We'd be shepherded into them during air raids. It was surreal. While the teachers couldn't give us regular lessons in the shelters, they told us stories as we huddled together. It was damp and smelly. But it felt safe.

While our school escaped damage, the old schoolhouse on the village green, less than a mile away, wasn't so lucky. It had survived over nine hundred years, a small wooden two-story building on heavily-treed church grounds. During the war, it was used for military drills by the Home Guard. The Home Guard consisted mainly of a bunch of doddering old men, clearly not eligible to fight, but determined nonetheless to do their thing for King and Country. So they played soldier, using the old schoolhouse as their HQ. Unfortunately for the rickety building, they used real guns and live ammunition. Once, during a mustering of the elderly troops, one of the Home Guard accidentally fired his rifle—boom!—and shot a hole in the ceiling. I think it is still there today.

Because he was 38 when the war started, my father was too old to go off and fight. Instead he volunteered with the fire department. While this naturally scared my mother, it was a great thrill for me, since he often brought home what my friends and I considered prized souvenirs—bits of German bombs. Sometimes they were curved brass, sometimes threaded. My shrapnel collection was larger than any of my friends. It was quite a source of pride.

Dad came home with the most exciting stories. And there were funny ones, too. Even as a volunteer, Dad slept at the firehouse. With the sound of the sirens, he and the other firemen would have to bolt out of bed and race to the fire. One day he sprang up when the siren blasted, forgetting he was in a lower bunk. He conked his head on the upper bunk, knocking himself out.

During the war my mother worked in a ladies' lingerie shop. Of course in wartime that meant strictly utility underwear, about as far from Victoria's Secret as you could get. Mom felt badly about leaving me alone during the day, and she brought me home postage stamps to make up for it. When I was five or six, she brought me home Hitler stamps, with swastikas on them. While I heard the name Hitler on the radio all the time, it seemed like a folk story. The stamps made him real.

While my father was home with me, my neighbor Wendy Saunders, who was a year younger than me, wasn't so lucky. Her father, Hedley Saunders, was injured at the battle of El Alamein. They put him in an armoured car at the battle site, but then the car was blown up and he was killed. Of course nobody told me this; I was four years old at the time. I overheard the adults discussing it. I remember feeling intensely sad for her and her mother. It was my first experience of death. That put a human face on the news we listened to on BBC broadcasts every day. I remember that we treated every one of those news broadcasts as a religious experience of fear and anticipation. I still remember the news announcer's name, Alvar Liddell, and the somber tones in which he announced the war's progress: "Gentlemen, today the news is grave. Singapore has fallen," ". . . the miraculous evacuation of Dunkirk," that kind of thing.

The combination of news, action and fear led me as a child to have many nightmares, some of which I still remember. My greatest fear of course was that the Germans would win the war, and for a very long time it seemed that everybody thought they would. Certainly in 1941 and 1942 my school friends and I thought a German victory was inevitable. The Germans had taken on a superman image. The BBC broadcasts were full of news of the British being beaten everywhere. It's true that the general population kept the traditional stiff upper lip, but there was a palpable sense of doom obvious even to a little boy like me.

Our British defenses at the time, in 1941 and 1942, seemed to consist only of "barrage balloons," which were tethered, blimp-like contraptions that were supposed to either stop the Germans from flying low, or alternatively to trap the German fighter planes by entangling them in the tether wires. There were apparently tens of thousands of these balloons moored all over England, but I never heard about any German fighter being knocked down by them.

Another British defense consisted of night-time searchlights and 'Ack-Ack,' or anti-aircraft, fire. This also seemed woefully inadequate, since I only ever saw one German plane actually get hit. It crashed in a local field and the farmer captured the crew with a pitchfork, much to the delight of the locals.

Our sense of doom wasn't improved by what we learned at school. I remember vividly our geography lessons, showing Europe covered with cross-hatching for the areas occupied by the Germans. Almost all of Europe was in German control, and the possibility that the allies could ever recapture countries in the center of the occupation zone, like Hungary and Czechoslovakia, seemed inconceivable.

My friends and I envisioned the Germans having control of our lives after they won. We thought there would be a German police force, but sought solace in believing that they would not intrude in our lives as long as we behaved ourselves. Hardly a nightmare scenario but not a pleasant one, either.

Our sense of German invincibility continued until the very end of the war, even though the air raids petered out in late 1942 and eventually ceased. All the way into 1945 we felt that the Germans would somehow spring a surprise and launch a major counteroffensive. It wasn't until V-E Day that our fears finally lifted.

The end of the war was celebrated in the biggest city near us, Birmingham, as it was in most of England, by bonfires. This was ironic since the Germans had made much of Birmingham a bonfire with their bombing attacks, leaving only a burned-out shell.

Chapter Two

Grammar School

By the end of World War II, Winston Churchill had been Prime Minister for almost the entire war. Certainly I could remember no prime minister *other* than him. With the Allied victory, it was unthinkable that the war hero Churchill would be voted out. But he was. In 1946, Churchill's Conservative Party was beaten by the Labour Party under Clement Atlee. (One of Churchill's famous witticisms came out of that political campaign. "Mr. Atlee is a very modest man," intoned Churchill, "and he has much to be modest about.")

As a nine-year-old, I had very little concept of how a great man like Churchill could be removed from office. But as it turns out, this was one of the luckiest breaks of my life. The Labour victory gave me the means, by way of an excellent education, to fight my way out of the lower middle class.

The Labour Party had a very different idea of how to run England than the Conservatives they'd beaten. The Conservatives reflected an upper-crust, status quo attitude. Labour, the party of socialists, believed in nationalizing big industry, providing free health and unemployment benefits—and of course the higher taxes to go along with those programs.

My parents were strong Labour Party supporters after the war, as were most of their working-class friends. Predictably, as they became a bit more financially comfortable, their views shaded to the right. I don't remember them ever discussing politics with me, although my father had some heated discussions with friends who were Liberals or Conservatives. I remember hearing Dad arguing for nationalization of certain major industries, and his friends arguing that without competition there would be inefficiency and no incentive to get things done better. At the time I agreed with my Dad out of familial loyalty. But as I grew up and saw what nationalization did to England, I realized his friends had been right.

Perhaps there has never been a person who didn't complain about taxes. But in England in the 1940's and 50's, there was a lot to complain about. The top level of income tax was 90%. Ninety percent! And there was a purchase tax on cars, watches, perfume and other "luxury" items of 110%.

Of course, the Labour Party didn't pocket the tax money. It went to pay for programs that wound up having a profound impact on me. Under the old Conservative regime, education past the age of fourteen had been a privilege for the upper and middle classes. Very, very few working-class children, no matter how talented or hard-working, had much chance of joining the educated elite. The socialists changed all that with the Education Act of 1944 (implemented in 1945). It provided that all children would take a comprehensive exam called the "Eleven Plus"—as the name suggests, at the age of eleven—and the children who scored in the top ten percent or so got to attend the previously private and privileged grammar schools of the country. The remaining children were shuffled off to "secondary schools" where education usually ended at age fifteen.

Of course the reason this new system affected me so greatly was that I was lucky enough to perform well on the Eleven Plus. As a result I got to attend the local grammar school, the Kings Norton Boys Grammar School. Had I been born even two years earlier, I would not have had that chance, and I cannot conceive how my life might have turned out.

I have often reflected on this because of a family tragedy. My mother had given birth to a stillborn boy a few years before I was born, a fact I was not to learn until I was about fifteen. One of my mother's friends casually brought it up in conversation. I was stunned and quite upset, more I think because Mom had never told me about it than the fact that I'd almost had a sibling. As soon as the friend left I asked Mom, "Why didn't you ever tell me about it?" She responded without betraying any emotion, "The topic never came up."

How, I wondered, would it have come up? I was unlikely at the dinner table to look up from my meat and two veg and say, "By the way, Mom, did you ever have a miscarriage?"

She attributed the miscarriage to "running for a train and straining myself," probably not the cause but a way for her to rationalize it.

In any case, had I been my almost-brother, only a couple of years older, my life experience would have been entirely different. No Eleven Plus, no grammar school, no university. The timing was that critical. I was in the first generation of English children to escape the class system through education, but just by a chronological hair's breadth.

Ironically, today the very Labour Party that championed educational opportunity for all in 1946 has turned its back on that system, judging it elitist because not everybody receives the same education! Ah, well.

At the Grammar School it became quickly obvious that the teachers, called "masters" in England, were somewhat nonplussed by the prospect of teaching boys like me from the lower classes. One master said to me, "We are not used to this class of boys." I like to think that what they lost in the upper classes they gained in native intelligence. I don't think *they* thought so!

All this talk of class—not school classes, but social classes—seems out of place in America. But as a child in England I was *very* conscious of it. Everybody was. My parents provided me with an idyllic childhood, but there is no question that the caste system in England soaked to the bones of our lives.

My father came from a working class family, albeit a relatively prosperous one. His mother died of pneumonia when he was only six years old. His father, at the time an insurance salesman, must have been quite prosperous because he owned a small house and was able to hire a housekeeper to look after my father, my aunt and another brother who was born mentally and physically handicapped and died at age six.

My grandfather subsequently married the housekeeper. She lived to the ripe old age of 96, and would have kept ticking even longer except for a tragicomic event. One day the termite infestation in her house got so bad that the floor collapsed under her, leaving her sitting, dazed and confused but physically unharmed, in her basement. She moved into an old folks' home while her house was repaired. While there, she caught pneumonia and died.

My father finished his education when he was fifteen. Possessed of a great natural curiosity, and coming of age at a tremendously exciting time of discovery and invention—he was born in 1902, and thus witnessed the birth of flight, radio, TV and virtually all of the scientific marvels of the twentieth century—he spent his first two or three years after school as an apprentice engineer/toolmaker. He voraciously read technical books, always expanding his considerable technical expertise. As I progressed through school and college, he was constantly interested in what I was learning, the elements of calculus and science that I would teach him as I learned them myself.

Before I was born, my father lost his job, along with just about everybody else during the Great Depression. He only succeeded in getting a new position as an engineer at Cadbury's, the chocolate factory, because he was attending night school in electronics and was able to convince the hiring folks that of the 1,600 applicants for the job, he was the one who would go the extra mile to learn and teach new technology.

It wouldn't be exaggerating to call him a technology nut. Even after he retired, he developed a little business fixing everything from watches to TVs, basically for free. Talented he was. A businessman he wasn't.

Because Dad was a kind and likeable person and obviously very skilled, I often wondered why his career at Cadbury's faltered. The answer—one I did not figure out until much later in my life—remains disturbing to me. He felt, as did many Brits of that time, that he was not good enough for opportunities for which he was clearly well-qualified. He felt that his accent labeled him as lower class, and that's where he ought to stay.

On top of that, I wonder whether the lower classes in England are born with an "anxiety gene" which, generation after generation, "keeps them in their place." I have no basis for believing this except for the experience of my father and my friends. Dad was promoted at least twice near the end of his career, the only times when he was in charge of other people. I can remember him being nearly frantic with anxiety for weeks on end. He would gradually realize he could not handle the situation, and he would ask for a demotion.

This anxiety also bore itself out in his music. As an amateur violinist and teacher, he was well-loved and quite competent. But when he was raised from violinist to chief violinist with a couple of local orchestras, if he had a solo to perform my mother and I learned not to speak to him the day of the concert. He would be edgy. He wouldn't eat anything. He couldn't sleep. He'd keep his hands constantly busy, working on this and that around the house in an anxious stream of motion. These anxiety attacks continued even after he discovered the drug Librium—a precursor to Valium—late in life.

My mother was very different from my father. I am pretty sure that whatever native intelligence I possess came from her.

Born in 1904, she grew up in a somewhat more prosperous family than my father, although her family was no more than middle class. As a child she showed unusual scholastic aptitude, and in fact won a scholarship to the George Dixon Grammar School. This was exceptionally unusual for a girl at that time. Tragically, before she could take advantage of the scholarship, she was diagnosed with tuberculosis. This was probably a false reading, as was often the case with tuberculosis then, but nonetheless it kept her out of school for a year. In those pre-antibiotic days, the treatment for tuberculosis was "fresh air." Missing school for a year destroyed her chance at a scholarship and effectively ended her education. She remained quietly bitter about it for the rest of her life.

I remember her as being an impatient, self-confident person, so very different from my father. Her frustrated intellect was buried in music. She had a remarkable gift of "absolute pitch," such that merely by listening to a note played on an instrument she could instantly identify what note it was. She was trained in singing and voice, but what was more remarkable was that she taught herself piano to the point where she could play almost any piece of music, in any key, without practice. I've never met anybody else with this ability.

She retained her confidence well into old age. In her seventies, she was playing piano accompaniment to dance recitals, from ballet to popular dance, at the Royal Albert Hall. In contrast to my father, she had virtually no nerves at all.

It wasn't until I was older and out on my own that I recognized a subtle tension between my parents. My father shirked increased responsibility in his career to avoid the pressure and anxiety that went with it. He was prepared to swap financial and personal aggrandizement for peace of mind. Mother had a very hard time dealing with what she must have regarded as his wimpish approach to life. I must have osmosized this attitude myself, because I guess I thought of my father as rather passive.

But on his deathbed, that changed. He said to me, "You must have wondered why, for so many years, I have let your mother make the major decisions in the family. Why I've seemed so indecisive myself."

Of course I *had* wondered about it, but I didn't say that.

"You see, in some small personal way I wanted her to be able to express her quick-wittedness and her decisiveness rather than suppress it."

I was floored. My estimation of him skyrocketed after that.

That conversation was one of the last we shared. It was 1980, and he had a few weeks earlier been felled by a blood clot near his spine. It left him paralyzed from the waist down. In the hospital, he seemed to be improving marginally. Mother and I were making plans for installing an elevator in their house, and generally making it wheelchair-accessible for Dad's return.

But that was not to be. Out of the blue, he died. I went over and was escorted into the hospital director's office. He proceeded to tell me, in very technical terms, how my father had died. He droned on and on, confident, I think, that I didn't understand anything he was saying. In summing up, he said, "So I'm sure you see that your father's death was in no way our fault, Mr. Walton."

I responded, "I should tell you that I am a professor with an adjunct appointment at a medical school. I fully understand what happened."

He blanched. The jig was up. He knew that I knew exactly what he'd been saying. It seems that a student nurse had, in giving my father an enema, ruptured the wall of his rectum and inadvertently killed him. The hospital director knew that this meant big trouble if I pursued it.

But I had no intention of doing so. The girl who did it was clearly mortified; it was a horrible mistake, and nothing more. On top of that, my father was elderly and his life, had it gone on, would not I think have brought him very much pleasure. I assured the very relieved hospital director that I understood what had happened, and that was that.

The common assumption is that once one elderly spouse dies, the other follows soon thereafter. Certainly my mother had never struck me, or anyone, as the type to soldier on alone. She had heart palpitations, was overweight, got little exercise, and ate a typically British fat and sugar laden diet. Perhaps more significantly, she didn't have a particularly optimistic personality. But her doctor told me, "I think attitude is seriously overestimated as an element of longevity. I've had the most pessimistic patients outlive the most optimistic ones."

Sure enough, my mother survived until the early 1990's, outliving my father by more than ten years.

XXXXXX

I often wonder how my parents would have fared given more educational opportunities. There is no doubt that attending Grammar School was my one route up and out of the lower classes. I did not realize this at the time, of course.

What I *did* recognize was an immediate—and lifelong—attraction to chemistry.

I'm not sure when I first became enthralled with chemistry. I have vague recollections of treasures in coloured bottles locked behind glass doors at my primary school, when I must have been about ten. There were all kinds of flasks and glassware, distillation and condensation equipment. Ooooh! I so much wanted to get my hands on these goodies. All I was told was that they were associated with "chemistry," and when I asked, "What's that?" I'd get blank looks in response. In spite of—or perhaps due to—this lack of concrete information, I imagined all of this stuff hooked together with coloured liquids bubbling through tubes.

Subconsciously, my life as a chemist had begun.

As my interest in chemistry grew, I read chemistry books as a hobby, and almost immediately wanted to create my own chemistry lab. I bought, begged and borrowed the equipment I needed to build a home chemistry lab and concoct my own experiments. *Everything* chemical fascinated me. I very much wanted to understand how things work, and it seemed to me that chemistry held the answers to those questions. Of course, fiddling with my chemistry lab also provided endless opportunities for practical jokes and all kinds of mischief.

I was always on the prowl for interesting chemicals, and would cadge whatever I could. I once bought some calcium carbide from a local chemist, with only vague ideas of what it was actually supposed to be used for. What I did know was that when you immerse calcium carbide in water it creates the gas acetylene, which is highly flammable. In the old days, before coal miners had battery-powered lights on their helmets, they used acetylene burner flames activated by calcium carbide, making a horrid job all the more dangerous. My plans for the calcium carbide were far less practical, and much more fun. In those days we still used fountain pens, and so our school desks were equipped with inkwells. I found that a small lump of calcium carbide in my classmates' inkwells would cause a miniature volcano effect, with bubbling ink oozing slowly across the desk top. Even better, it smelled ruddy awful.

I wasn't the only one at school who cottoned on to the inkwell volcano trick. One day, one of my classmates set an unsuspecting victim's inkwell oozing. The master teaching our class was furious. "Stand up!" he demanded. We did. He clasped his hands behind his back, and paced back and forth at the front of the standing assemblage. "All right," he said. "Who did it?"

No response.

"That's fine," he said. "You will all remain standing, all night if necessary, until the culprit comes forward."

My friend Lamb (we all used last names in school) slowly raised his hand, as though to confess. The rest of us were shocked at his treachery.

"Sir, I cannot tell a lie," Lamb said, invoking George Washington. "Banner did it."

Everybody, including the master, collapsed in laughter. We all got off scot free.

xxxxxx

Many of my classmates were as fascinated with chemistry as I was. We all knew how to make run-of-the-mill stink bombs with hydrogen sulfide. But we quickly learned how to make much more subtle organic compounds that had indescribably gruesome odours. I can remember at least one occasion when the grammar school had to be evacuated because someone had flushed one of these substances down a drain, polluting every classroom in the school. It was just as effective at clearing out the building as pulling a fire alarm would have been—but it was a lot more fun.

Once, when we were in Sixth Form (the equivalent of being high school seniors in America), we were studying the interesting properties of phosphorus. It's an element rarely seen in its elemental form, and it is very unstable in the air. It can easily burst into flames, spontaneously forming choking and poisonous phosphorus oxide clouds. It is not much better in water; a chunk of phosphorus placed in water explodes and forms phosphoric acid. The grammar school's phosphorus sample was tiny. It probably only weighed about ten grams. It was kept under the organic solvent toluene, so that it would be stable.

Our chemistry master, Mr. Moore, was a wonderful teacher, but we felt he lacked a certain sense of humour. That made him a prime target for chemical jokesters. One of my classmates succeeded in spiriting some of the phosphorus out from under the toluene and spreading it on the floor behind Mr. Moore's desk right before our chemistry class.

About halfway through the class, one of the students raised his hand.

"Yes, what is it?" demanded Mr. Moore.

"Well, Sir," the student said, "I think your shoes are on fire."

Sure enough, there was a significant amount of smoke billowing from Mr. Moore's shoes. He took one horrified look at his feet, ran over to a fire bucket full of water, leapt in and stood calf-deep in water . . . much to the class's delight.

xxxxxx

When I was about sixteen, I learned the secret of making silver acetylide. This marvelous stuff was perfectly safe when wet, but when dry, it detonated on contact. What fun!

I made the silver acetylide by bubbling acetylene gas into a solution of silver nitrate, the stuff used in photographic solutions. I painted blobs of the wet, gray solution on the kitchen step and watched anxiously for it to dry. When it finally looked ready, I dragged my dog Bob over to the step, and he innocently stepped on the "paint." BANG! Bob leapt into the air with a start. Clearly if it startled Bob I could hardly wait to try it on a human guinea pig. Specifically, my father.

Our garage floor was paved with flagstones, an ideal substrate for the silver acetylide. My mother and I waited anxiously in the kitchen for Dad to come home. Sure enough, he walked in *very* excited, telling my mother, "Hilda! The garage is full of static electricity!"

Mom and I exchanged a conspiratorial smile. In fact, she told me afterwards that when she and my father were dating, he'd actually come up with a very similar experiment. He had made two damp spots on the rug—presumably with water—and connected them to an electrical device. When Dad's dog walked onto the wire, it jumped about a foot in the air. By standing on the damp spots and the wire, the unsuspecting mutt completed a circuit and got a minor shock. So in at least this one aspect, I was a chip off the old block.

On one occasion, when I was about thirteen, I wanted to make an acetone extraction of chlorophyll. To do that I needed to harvest some leaves. Armed with a lot of evergreen needles, a large pot, and the appropriate chemicals, I used my mother's kitchen stove to heat the mixture. POOF! Suddenly the pot was empty, and the ceiling was plastered with evergreen leaves.

Fortunately my indulgent parents didn't put a stop to my experiments. When I was about fourteen I wanted to get my hands on some of the solvent benzene. It wasn't possible to get benzene from any conventional source, which made it even more desirable.

Mothballs, on the other hand, were very easy to come by. My mother always had them in ample supply. Mothballs were made of paradichlorobenzene, and I seized on the idea of extracting the benzene from those mothballs. I knew this wouldn't be easy and in fact there wasn't a chemical reaction I knew of that would work. But I was absolutely obsessed with the idea. I turned it over and over in my mind, figuring that if I could catalytically "crack" the dichlorobenzene by hydrogenization, the product would be benzene and hydrochloric acid.

Tremendously excited, I put together the necessary equipment in our kitchen while Mom and Dad listened to the radio in the living room. I hooked up a distillation apparatus that passed a mixture of "mothball" vapor and hydrogen over heated granulated tin, which acted as a catalyst.

BABOOM! Instead of benzene I got a jolly great explosion. Sizable chunks of glass flew about the kitchen, embedding themselves in the cabinets. When my parents heard the explosion, Dad exclaimed, "What was *that?*" to which my mother calmly replied, "Oh, it's just Alan experimenting again."

But she was sufficiently concerned to walk over and open the kitchen door to survey the damage. Her mouth fell open. As her eyes swept around the room, the light glancing off the shards of glass gleaming menacingly from her cabinets, she was speechless—a rare occurrence! After a moment in shock, she rushed in to make sure I wasn't hurt.

Perhaps my parents' relief at finding me intact lessened their anger, although Mother was *not* pleased about her defenestrated kitchen cabinets.

Despite the occasional kitchen mishaps, explosions continued to fascinate me. I was delighted to learn that our family doctor had been an explosives expert during the War. From him I learned one of my best tricks, which was to make an incendiary bomb by mixing sugar and potassium nitrite with a drop of sulfuric acid.

While my extracurricular science experiments were hardly constructive, my enduring fascination with chemistry—and later biological chemistry—carried me through grammar school, national examinations and university, and in fact through my life ever since. Although I didn't recognize it at the time, the defining question of my professional life was triggered the very first time I gazed into those glass-fronted cases at primary school:

"How does it work?"

Chapter Three

Music, Music, Music

Chemistry has been my abiding passion most of my life. But until I was about ten, when anyone asked: "What do you want to be when you grow up?" I'd promptly answer: "A professional musician."

My parents were both amateur musicians. My mother was a talented pianist, and my father a part-time violin teacher. He bought me a half-sized violin when I was about four, probably with the vision that I might turn out to be a modern Mozart. It wasn't meant to be. *Skree! Skree! Skree!* I scratched out whatever I could on the violin, but it wasn't much.

I did a bit better with other instruments, although my early dreams of being a professional musician ended abruptly with my first recital, when I was ten. It was hellacious! I sweated through Bach's *Minuet in G,* more terrified than I've ever been since. That might have been the end of music for me, except that a couple of years later I got the chance to play with an amateur orchestra. It wasn't because I had suddenly found a talent for violin. Instead, Dad pointed out, "They're trying to put a youth orchestra together in town. There aren't enough musicians to fill all the spots." With the talent pool pretty much dry, even *my* odds were good. Then I learned that there were no auditions, and I became a sure thing.

The challenge of playing music with other children—awful though it may have sounded to the audience!—brought a dimension to my life that had been missing. While the idea of cooperation and pulling together was something that other children found routine, as an only child it was a novel experience for me.

When I got to Nottingham University, music suddenly became fun. The only music I had ever played was classical, due to my father's influence. He considered jazz a cacophony, and it wasn't until I went away to

University that I learned the excitement of all kinds of music. In my second week at Nottingham, the English Society had a party where I was exposed to two wonders that were to heavily influence my University career: beer, and New Orleans jazz via the Mercier jazz band. The band had seven pieces—trumpet, clarinet, trombone, piano, banjo, tuba and drums. The players were all students at either Nottingham or the local arts college. They practiced two or three times a week at the University during lunchtime, and a hundred or so of us would take sandwiches and listen to them. I thought they were just fabulous. I watched in awe as they played *without sheet music*, extemporizing everything. I didn't know at the time about chord progressions, but soon learned the form and shape of music played without written "instructions."

I didn't, however, turn my back on classical music. I joined the University Symphony Orchestra, a big thrill for me. It had a hundred or so musicians, and the quality of playing was far superior to anything I had previously experienced.

I played the G-trombone if the music called for it, and violin if it didn't, my violin playing having improved somewhat since my first recital. The G-trombone, in case you aren't familiar with it, is a part of every big orchestra, but there aren't many people who play it. I have no idea what was in my mind when I took it up, because it wasn't easy.

The secret to playing the trombone—as is true with all brass instruments —lies in two elements: the way you move your lips, and the way you manipulate the mouthpiece. You can't just give a trombone a straight "blow job" and produce any sound. Instead, you have to make a kind of a farting sound with your lips into the mouthpiece. This sound is amplified through the instrument and comes out as a note depending on how you set the slide (the big, paperclip-like tubes which make the trombone distinctive).

You can produce several octaves' worth of notes by holding the slide in the same position while tightening your lips and moving their position slightly against the mouthpiece. As if all that weren't difficult enough, you control the quality or tone of the note you produce by vibrating your lips against the mouthpiece. So the bigger the mouthpiece, the easier a brass instrument is to play. The French Horn is the most difficult brass instrument of all, because it's got the smallest mouthpiece. The slightest mispositioning of your lips, and you produce a sound more like an animal wailing its last breath than a musical note!

The G-trombone I played also had the distinction of being one of the largest instruments in the orchestra. It measures up to six feet in length at full extension (although G-trombones today are a bit more compact than they were then).

While the size of the trombone wasn't a problem when we were playing on a regular stage, things were far more comical when I played with the "light orchestra," or pit orchestra, for University performances of the Gilbert and Sullivan operettas which were very popular at the time. The musical scores for these operettas always called for a more or less full orchestra, including a bass trombone. The trouble was that the available space for a pit orchestra is very small, and not well suited for a G-trombone. On one occasion during a concert performance, I managed to catch a stand holding cymbals, crashing them to the floor. On another, I was watching the action on stage and burst out laughing at a stunt just as I put the trombone's mouthpiece to my mouth, with a somewhat alarming result. Fortunately the audience seemed to think this trombonery was part of the plot.

This was hardly my most embarrassing trombone experience. *That* encomium goes to a time when we were rehearsing for Gilbert and Sullivan's operetta *Iolanthe*. As it happens bass trombone players are rare and highly sought-after, the rarity coming from the fact that most pieces of music contain very few notes for the bass trombone. *Iolanthe* was no exception.

One part requires about a thousand bars' rest and then a brass solo entry, where the bass trombonist is required to play half a dozen notes somewhere near the top end of the range. This is very, very difficult, because the higher the note you have to produce, the more likely it is that you'll get it wrong.

The long waiting period created such excitement and stress that when I actually got to the part where I was supposed to play, my lips had a mind of their own. I attempted the entry a couple of times, at which point the conductor stopped the orchestra and said, "Walton, what is the problem?"

"Well, Sir," I responded, "I blow the right note in, but the wrong one keeps coming out!"

Ideally the audience at the actual performances of *Iolanthe* didn't notice that while the rest of the brass section played the fanfare, the bass trombone's role was taken over by the piano.

My favorite classical mishap at University wasn't one of my own, and didn't even have to do with playing an instrument. The timpani (kettledrum) player with the University Symphony was lucky enough to attend the re-opening of the Festival Hall in Salzburg, Austria, in 1957. This was at a time when Austria had been re-unified, after the Russians had returned Vienna to the republic of Austria following the War.

He brought with him some chocolates to eat during the performance. More specifically, they were Maltesers, the British equivalent of malted milk balls. They were one of the few things that weren't rationed in England at the time, during a period when the availability of virtually everything, including currency, was strictly controlled.

He found that the Festival Hall was newly furbished, but had not yet been equipped with a carpet, whether for acoustic or economic reasons. His seat was near the back of the auditorium, which had a series of platform steps sloping towards the orchestra. He sat enraptured by the music and happily munching his Maltesers.

The music built from a fine crescendo to a climax which involved a short, silent pause. At that moment, he dropped the Maltesers, and they cascaded down the steps—Plink! Plink! Plink! Plink! Plink!, echoing as they tumbled toward the orchestra pit. The noise, amplified by the hall's marvelous acoustics, made the whole audience turn around and scan the crowd for the idiot of the moment. My friend had the marvelous presence of mind to do the same. He turned and gazed in scorn at a young man sitting alone at the back of the hall!

xxxxxx

At University, it quickly became clear to me that playing only the trombone and the violin limited my musical options. I realized that if I could learn to play the string bass, I would be able to play in modern and traditional jazz groups, dance bands, rock groups, as well as orchestras of all sizes. It took me two years to learn to play the bass. While I had taken trombone lessons —fat lot of good they did me!—I got hold of a book about the string bass and taught it to myself.

I gradually worked my way into all different kinds of groups, from the Nottingham Philharmonic and Nottingham University Symphony Orchestras to the rock and jazz groups. I found that the string bass had a very attractive feature in common with the bass trombone: because not many people knew how to play it, I got to sit in with groups long before I was really proficient. Fortunately for me, the bass is fairly easy to fake with jazz groups!

My friend Terry Willis and I quickly realized that we could make a fair bit of money if we put together a musical group of our own, because many University functions required a dance band.

Of course nothing is as easy as it seems when you are making big plans with your buddies over a couple of beers. We started out doing some really terrible gigs. While weddings were usually fun, bars were the worst. On one occasion, Terry Willis went out with another guy to play a gig at a local bar, with Terry on piano and the other guy on drums. (We didn't usually go out with fewer than four players, but only Terry and this other guy could make it that night. He must have really needed the money.) Pianos at bars were normally pretty lousy, and no matter how well he played Terry couldn't get the crowd's attention. So as a gag, in the middle of one tune, Terry started playing the piano with his elbows. At the end of the piece, a drunkard staggered over to Terry and said, "I think you were out of tune in that piece."

Terry replied, "Oh, that wasn't me. It was the drummer." At which point the drunkard started berating the drummer for his lack of musical talent!

Even as our bands started doing well, we didn't have the money for top-shelf equipment, and this sometimes created problems. At one point our guitarist, Ray Holland, had been having trouble with his guitar amplifier. It was probably a bad connection. With a big University bash coming up, Terry Willis and I implored Ray to make sure that his amplifier was overhauled. He assured us that he had spent many hours checking everything out, and that it would be fine.

Came the big night and the band was setting up. The ballroom filled with dancers, and we were poised to launch into our big rock intro, a variation of Bill Haley's "Rock Around the Clock." Ray plugged in his guitar and to our horror, POOF, the lights went out on stage. They also went out in the ballroom, the University, and the entire city of Nottingham. By incredible coincidence, the City suffered its one and only power failure in living memory at the instant that Ray Holland plugged his guitar into his amplifier!

We were finding that by organizing some of our skilled musician classmates and slightly undercutting the musician's union rates, we quickly dominated the commercial music scene at the University. And as our reputation grew we played gigs within a fifty-mile radius of Nottingham. We even organized a couple of other bands so that we could play multiple gigs at the same time. As our fame spread, our prices increased. We played jazz as well as early rock improvisation, and some of our players went on to become professional musicians. For instance, there was Johnny Ayers, our lead saxophone/clarinet player for jazz gigs. He was one of the cleverest jazz players that I ever met. He certainly wasn't remarkable looking, a rather plain fellow of medium height. And he was rather quiet. But what a musical genius! He could listen to a group, and immediately pull apart what everybody was playing. He could formulate all of the notes of any chord, in any key, in his mind. He could listen to a tune once, and then play it on the clarinet or saxophone. Just an astonishing talent.

Ray Holland, our lead guitarist (with the amplifier problems), could alternate between a mean rock guitar and a Segovia style classical guitar. He also went on to play professionally.

A professional music career even became a possibility for me, although not in the way I'd imagined. As the band became better known, it became clear that we needed vocalists. Since I was the only person in the group who could carry a tune, and we had a rhythm guitar player who could take over my bass duties, it fell on me to be the singer.

There was only once that I can remember that we bombed terribly with me singing the lead. We had a gig at a local dance hall where we played for a bunch of "Teddy Boys," as bikers were known then, and their dates. We prided ourselves on our fairly wide repertoire of music, but all they wanted to hear were cha-chas. Our selection of cha-chas was pretty negligible, and when we ran through it early in the evening, we invented a few and played those at least twice. It's a wonder people didn't throw things. Not too concerned with our musical egos, the proprietor shook his head at the end of the evening and told us, "You were pretty bloody unpopular."

Aside from that hiccup, we were steadily getting more and more popular. Our gigs were drawing bigger and bigger crowds. We played rock concerts at factories in the area at lunch hour, and sometimes had several hundred factory girls show up for our performances. There were even occasions when the police had to hold back the crowds. What an incredible taste of power that gave us!

The band was taking up more and more of my time. It wasn't unusual for me to work on my Ph.D. research from 9:30 in the morning until six at night, rush home, grab my musical equipment, and play a gig until well after midnight.

Our music careers were snowballing. We auditioned for a recording contract, which prompted the University newspaper to do an article about us. The article bandied about the possibility of our recording career, and featured pictures of the band. One photograph showed me, a cowlick at my forehead and crooning into a microphone, with the caption "Elvis Walton" (to this day my daughter Kimm occasionally pulls out this newspaper clipping for the amusement of family and friends).

In fact, that newspaper article brought about the most serious dilemma of my life up to that time. Professor Dan Eley, the esteemed Head of the

Chemistry Department at Nottingham, called me into his office. He thrust a copy of the newspaper toward me, and said, "Walton, either you are going to do this silly rock-and-roll stuff, or you are going to get your Ph.D. But you are not going to do both."

My life fell apart at that moment. I loved music, I craved the thrill of playing in front of increasingly adoring crowds, and I couldn't fathom giving that up. But doing research was not just an honor. It was a dream come true. I had fought through eighteen years of education, and clearly I couldn't turn my back on that.

Eley stared at me expectantly. Faltering, I stammered: "Could you give me twenty-four hours to think about it, Sir?"

In retrospect, I can see that my answer could have had me cashiered on the spot. Eley was a very serious man with a very serious commitment to the intellectual pursuit of science. But by some stroke of luck, although I had not previously detected a sense of humour in him, Eley glowered at me for only a moment before bursting into laughter. He thought I was joking!

Thus ended my rock-and-roll career.

Chapter Four

Used Cars and Rolls Royces

The music gigs I played in college were a blast in every way. But to get to gigs, I had to have a car. My need for wheels inadvertently led me to another money-making venture—selling used cars.

As the 1950s came to a close, the car industry in post-War England was only just returning to normal. In fact, for a year or two after the war it wasn't possible to buy a new vehicle at all. In the late 1940's, the government strictly allocated new car sales. If you were one of the lucky few chosen for the opportunity to buy one, you could turn around and sell the car you bought for double what you paid.

By the time I was a student at Nottingham in the late 1950's, many pre- and post-War cars were available second-hand. There was a used car dealership in town, and the centerpiece of the showroom floor was a red, sleek, sexy Lloyd sports car. Now I *could* have bought a sensible Austin, but from the moment I first set eyes on the Lloyd, I knew I had to have it.

I scraped together every penny of my savings and went to the showroom with a buddy of mine. When he saw the beautiful Lloyd, he was just as excited as I was. I asked the dealer, "Could you start it for me?"

"Wellllll—all right," he said hesitantly. His reluctance should have immediately clued me in to the condition of the car. But lust has a way of clouding one's common sense.

In fact, he did get the engine to turn over a couple of times. Then it started and belched out huge amounts of white smoke. "Oh, that," the dealer said, a nervous giggle in his voice. "It only does that because it has very advanced engineering, a supercharged two-stroke, two-cylinder engine which requires an oil-gasoline mixture to work effectively."

"But where is the smoke coming from?" I asked, tears pouring from my eyes because of the fumes.

"Of course, the smoke comes from the oil burning off in the gasoline. When she's warmed up, the smoke goes away."

This made sense to me. "And she's got less than fifteen thousand miles on her, she's practically new," urged the dealer.

I was sold.

Needless to say, on the way home the smoke didn't go away. It got worse. And I could have lived with that, I guess, but then the car broke down. It was the first of many adventures with the Lloyd.

On the plus side, owning the Lloyd did give me a chance to learn more about engines than I ever thought I'd need to know.

One of the main problems with the car was that it had an aluminum engine block that tended to warp and blow its cylinder head gasket. Since the manufacturer had (wisely) discontinued the Lloyd, I had a heck of a time getting new gaskets for it. A mechanic told me that if the cylinder head was machined properly, I could get away with a gasket made of cardboard, using an appropriate metal "glue."

I tried this repair, but the car petulantly refused to start. My Dad, an engineer who could probably have built a working engine out of popsicle sticks, volunteered to help me out. So he and my Mom drove up to the University. I took them over to my garaged Lloyd, and explained what I thought was wrong with it.

"Oh, no," Dad said. "I don't think the gasket is your problem. Why don't you take out the spark plugs. I'll turn over the engine and you can listen for water percolated by the pistons. That'll tell you if the head is really leaking."

Mother stood by and watched this "boy stuff" with some disdain, as I removed the plugs and placed my ear next to the spark plug holes in the cylinder head. Dad got into the driver's seat and pressed the starter button. I don't know what he thought would happen, but the next thing I knew, two jets of rusty water shot alternately out of the spark plug holes, hitting me in the face and running down my formerly white shirt.

Since the raised hood blocked Dad from actually seeing what was happening, he kept the engine turning. I was too astonished to move. Mother collapsed with laughter. She told me later that it reminded her of a Laurel and Hardy show.

With Dad's help I did eventually get the Lloyd running, at least sporadically. It was quite a head-turner on campus. Not only was it beautiful to look at, it made quite an attention-getting roar. But of course with all of its quirks it was hardly the dream car it seemed to be. I used to call it a sheep in wolf's clothing. Even when I had it running on a fairly regular basis, I am quite sure that some of the exhaust was draining into the passenger compartment. This wasn't a problem when the top was down, but when it rained and I had to replace the top, I'd nearly suffocate.

I never knew when I hopped into the driver's seat whether the car would cooperate or not. And even when it did, making it all the way to campus was a fifty-fifty proposition. For most of the time I owned the Lloyd I lived in Beeston, a mile or so from the University and an easy walk. Whenever I was late to work, the lab staff would joke, with some merit, "Walton must be driving today!"

On one bright sunny day, I was driving back from the lab to my flat. The weather was delightful and the Lloyd was running well, for a change. As I drove down the main street I could see many heads turn in admiration. It made all of my travails with the car worth it.

Suddenly there was a loud explosion. A jet of steam shot twenty feet in the air, and I came to a screeching halt.

I knew immediately what the problem was. The cylinder head gasket had blown again, forcing supercharged exhaust into the radiator. The header tank had exploded under the pressure, creating a fair imitation of Old Faithful.

I'd never seen anything like it. None of the passers-by had, either. Within moments I leapt out of the car, and a crowd stood around me gaping at this extraordinary phenomenon. Half a dozen of them were very helpful. They helped push the car to a nearby repair garage.

The combination of a constantly broken-down car and a lack of funds to pay to have it fixed made me, of necessity, a pretty fair mechanic. I soon decided that I could make money with that knowledge.

A little digging unearthed the fact that the Nottingham campus would be a tremendous market for used cars. Of the twenty-five hundred students on campus in the late 50's, only about fifty owned cars. Having your own car was a *big* status symbol. And it didn't matter what kind of car it was. Most of the fifty cars on campus were pre-War Austin 7's or Morris 8's. Those would run no more than 50,000 miles or so before the engine had to be replaced.

The Austin 7 was known for having hideous brakes and a weak electrical system. In fact, a friend of mine was driving his Austin 7 from Nottingham to London one night when, as usual, the lights blinked out. Instead of stopping to fix the problem, he continued to drive with one hand on the wheel and his head under the dashboard, digging around with his other hand for the problem. Not realizing where he was going, he drove under the back of a truck, shearing off the windshield and the top of the car. Fortunately, he wasn't hurt. He subsequently cut the jagged remaining top from the Austin, and used it as an open touring car after that.

Another problem with both the Austin and the Morris was that they had very weak door latches. One friend was making a left hand turn in an old clunker when the door flew open and he catapulted into a ditch, scrambling up just in time to see his car continue down the road without him.

The old cars also had a lot of "play" in the steering wheel. You could sometimes turn the steering wheel almost halfway around before the wheels responded at all.

The bottom line was that even students who owned cars would soon be in the market for a replacement. And I figured that if I could come up with really *cheap* cars—and I mean really cheap—I could make a bit of cash. Other than the American students who were dripping in cash, nobody could really afford much more than fifty pounds for a car (about fifteen hundred dollars today). The lack of money made students very imaginative when it came to transportation. A friend of mine needed a pickup truck, but couldn't get his hands on one. What he *did* locate was a very old Rolls Royce sedan. He ingeniously hacked off the back half of the passenger compartment, and rigged it up like a flatbed truck. It suited him fine, but it didn't suit the Rolls Royce Company. Apparently a Rolls Royce owner had spotted my friend's bifurcated Rolls, and it had put his nose out of joint seeing a Rolls Beverly-Hillbillyized like this. Shortly thereafter, my friend received a missive from Rolls Royce, discreetly inquiring whether he might be persuaded to sell them his—ahem—Rolls Royce. He did, and made a handsome profit from it.

A little bit more research uncovered the fact that most of the used car dealers bought their cars at a local auction. There wasn't any reason I couldn't do the same thing. So with an amazing stroke of luck—I actually found some poor sucker to buy my beautiful but temperamental Lloyd—I was in the used car business.

I would generally buy old Morris 8's for about 25 pounds and sell them for 50 or 60 pounds. I often owned as many as three cars at a time. For a while things went well, and I was able to buy an MG sedan to get me around to music gigs.

But I made a mistake. I sold one of these cars to a friend. He drove it for about a year, put ten thousand trouble-free miles on it—much more than I'd thought the dear old thing could manage—before the rear axle fell out. He was livid. I pointed out to him that considering how many miles he'd squeezed out of it, I'd given him a terrific deal. But he was not to be appeased. In the end, I agreed that I'd buy another junker Morris 8, and he could take the differential gear free of charge as long as he would disassemble the rest of the car and help me sell the pieces.

So I bought a non-working Morris 8 for ten pounds, towed it to one of the more secluded University parking lots, and we proceeded to dismantle it.

The following week I got a call from the Vice Chancellor of the University. "Walton," he said, "I hear you are operating a used car parts business from one of the University's parking lots. Would you kindly vacate the space immediately!"

I wound up piling all of the remaining parts into the shell of the Morris. I towed it to a scrap yard. The scrap yard owner gave it a once-over, looked at me, and said, "I'll give you ten pounds for the lot." That's exactly what I'd paid for it.

And I was out of the used car business once and for all.

Chapter Five

Alan and the Chocolate Factory

During two of my summers at University I worked at Cadbury's, the chocolate factory. My father had been an engineer there for many years, and he set up my first job, which was in the complaint department. It was a lot of fun being the first one to see the complaint letters that came in. I guess part of why I liked that job so much was that I never had to *respond* to the complaints, I just passed them along to the appropriate people. Considering the huge volume of chocolate that Cadbury's sold, there were really very few complaints. And mostly they were trivial, reflecting the exigencies of chocolate manufacturing at the time. It wasn't at all like the Willy Wonka chocolate factory!

For instance, many of our complaint letters came from people who'd found a hair in a chocolate bar. This was unavoidable in the 1950's, since chocolate production hadn't yet become sufficiently automated to keep production workers away from the chocolate as it was manufactured. Even though everybody wore protective hats, an occasional stray hair would find its way into the chocolate. Another problem was fingerprints on individual chocolates. Since chocolates at that time were hand-packed, when the weather was warm it was almost inevitable that some of them would bear a telltale fingerprint or two.

In these cases the customer would get a letter of apology and a gift of a one-pound tin of Cadbury's premium chocolates, called "Roses" chocolates.

In a perverse way, when you work in complaints, you secretly pine for a real whopper. I did hear through the grapevine that some years previously, a mouse—or more appropriately, some mouse parts—had been found in some chocolate.

The most curious case during my stint consisted of a letter of complaint and a sample of a chocolate bar. I could plainly see a hexagonal metal machine nut embedded in the chocolate. It had clearly fallen from a machine and molded almost perfectly into the bar. Also in evidence was an unmistakable set of tooth marks around the edge of the metal nut. The letter read as follows:

> Respected Sirs,
>
> I am an elderly lady who has enjoyed your chocolate bars for many years. I recently purchased one of your bars before entering a cinema and was busily eating same in the darkness during the movie. Now I understand that your chocolate is clearly marked *nut* chocolate, but I was very surprised and anguished when I bit into a *metal nut* and broke my false teeth. As a result, my family has advised me to sue Cadbury's and seek appropriate financial remuneration. I look forward to your favorable response.
>
> Yours faithfully,

It seemed to me that this was a fairly serious situation, requiring the input of my supervisor. He read the letter and examined the evidence, and as I watched him for a reaction, he started to laugh, slowly at first but then guffawing. The laughter rippled through the rest of the office as the story circulated.

After he calmed down, he taught me my very first lesson in business: namely, aim low. He said, "I doubt that she broke her teeth. But she has sent us the evidence. If she does sue, there is nothing we can do about it. So"— he paused dramatically—"Send her *two* boxes of Roses chocolates."

We never heard from her again.

The following summer I got the job at Cadbury's that I'd been coveting: packing Roses chocolates at the factory on the night shift at the factory in Bourneville, a suburb of Birmingham. For a start, it was very well paying. As I was being trained, my delight with the job turned to euphoria when I heard the response to my question, "How much chocolate are we allowed to eat?"

"As much as you want," the supervisor shrugged.

I couldn't *believe* Cadbury's generosity. All the chocolate I wanted! I had visions of never buying a single grocery for the entire summer.

Of course, I quickly learned Cadbury's wisdom in settling on this policy. In the first two days, you'd eat about five pounds of chocolates, and not be able to force down even one more the rest of the time you worked on the line! That's what happened to most of the college students I worked with.

However, there was this one remarkable fellow, an Indian undergrad who was studying at Birmingham University. He held a regular day job, slept four hours or so and then worked the night shift at Cadbury's. Instead of indiscriminately gobbling down chocolates as the rest of us greedy buggers did, he carefully assembled his "dinner" every night from the components of the chocolate centers. He'd have dried fruit and nuts for the main course and various sweet fruit mixtures for dessert, to give himself a somewhat balanced diet.

At the Bourneville factory I was one of about a dozen packers. The various lines of chocolates would come past us on an assembly line. The idea was to pack one of these lines into little paper cups, and then into boxes, in two layers. About two thirds of the people working the line were regulars, mainly Irish immigrants, and the rest were college students on summer break.

At that time, Cadbury's was not air-conditioned (which is why they had the problem with fingerprints on chocolates). If it got too warm, they couldn't run the chocolate factory. Apparently they'd calculated some years before that the lost production revenues from plant shutdowns were less than the cost of air-conditioning the factory.

But the summer I worked the line, 1955, probably changed their minds. For three out of the eight weeks I was there, it was sufficiently warm—80 degrees or so—that we would report for work, wait for a couple of hours or so to see whether it would cool off enough to start the production line, and then we'd be sent home. While we waited, we'd play bridge, then and now my favourite card game.

During this time I met one of the most remarkable people of my life. I'll call him Turton. I knew of him long before I met him at Cadbury's. He was a legend at the grammar school we both attended, King's Norton Grammar School for Boys, where he'd graduated about five years before me. Our physics master had always chastised my classmates and me for being slower than the famed Turton.

Stories about him had been handed down over the intervening years. In one of them, he'd been gazing out the window during a physics class. The physics master, sure that Turton had not been paying attention, asked him some obtuse question, confident that even if Turton had been paying close attention he wouldn't be able to answer. Turton strode to the front of the class, worked through the mathematics on the board, underlined the answer, and sat down, never uttering a word.

No wonder we poor mortals could not measure up to the great Turton!

By the time we were both working on the production line at Cadbury's, Turton was a graduate student at Birmingham University. He was studying for his Ph.D. in mathematics, although he was on leave that particular summer. While a Ph.D. in mathematics may not sound like a big deal, at that time the number of individuals receiving doctorates in mathematics from *any* British university hovered near zero. Even professors of mathematics at many universities didn't have a Ph.D. in the subject. As an undergraduate Turton had earned a first class honors degree, which is very rare. He'd probably been the first one in ten years to receive one.

The story of his degree exam was in itself probably apocryphal. The final exam consisted of twelve questions which required a very deep understanding of mathematics and its underlying meaning. It was probably necessary to answer correctly four of these problems in order to get a first class degree, and most people couldn't answer even one. Turton answered eleven correctly. For the twelfth, he pointed out an error in the problem, modified it, and answered his modified question correctly.

True or not, it was obvious that Turton's grammar school genius had not deserted him by the time I sat opposite him on the production line at Cadbury's.

The most immediately obvious facet of his genius was his photographic memory, particularly relating to one-line jokes. He could go on for the eight-hour shift for five nights a week, never repeating the same joke. It was quite funny at first but rather annoying in the long run. To divert him from this incessant joke-telling, I persuaded him to talk about himself, and with some reluctance he told me the following story.

After graduating with his undergraduate degree, he had gone into the Ph.D. program, where he was given a classical problem. It had been known to mathematicians for over a hundred years but it had gone unsolved. Within nine months, Turton had developed a new form of mathematics to solve the problem!

While this remarkable achievement was undoubtedly worthy of a Ph.D., there were clouds on the horizon. It seems that during the time Turton had been working on the problem, there was a series of minor thefts at Birmingham University, where he was studying. One young lady had her briefcase stolen, containing all of her lecture notes for the semester. Frantic, she posted a notice indicating that the thief could keep the briefcase, but would he or she please, please, please return her lecture and exam notes.

She indeed got the notes back, in a brown paper wrapper that had obviously been recycled. She saw that the previous addressee was none other than Mr. Turton!

It turns out that Turton was a kleptomaniac.

He was hauled into court on theft charges. The defense argued that this was one of Britain's greatest young scientists, and if he was slightly impaired psychologically, the treatment ought to be psychoanalysis rather than criminal punishment. The court agreed that if Turton were suspended from university and underwent therapy, he would receive a suspended sentence.

It was during this suspension that I met Turton. Apart from his fascinating mathematical genius, there was really nothing to commend him. He was truly creepy. Indeed, after his advances to one young lady were snubbed, he didn't shave or bathe for a month.

While he was working on the line at Cadbury's he was still dabbling in mathematics. But he mainly spent his time outside the factory listening to the radio—where he learned his one-line jokes, presumably—and going to the court-ordered psychotherapy sessions. He regarded his psychologist as a moron whose every move was so obvious and limited that "I could be a better psychologist myself." I had no doubt he could.

But all of this paled in comparison to his gigantic intellect. It was easy to overlook his many failings just to hear him talk about how he solved previously-unapproachable mathematical problems. His *modus operandi* was absolutely remarkable. He told me that he would examine the problem from all known mathematical approaches. When those didn't work, he would then try to understand the limitations of existing methods, think about how these might be overcome, and then he would put the problem away for months on end.

Then, suddenly, when he was absorbed in something else—a walk, perhaps, or sitting listening to the radio at home—he would get a brainwave and develop a new form of mathematics. Turton made me recognize that some scientific problems are not solved by traditional methods, but rather involve a major "eureka!" development that could not be derived from the previous state of that particular scientific discipline.

In fact, many years later I was at a conference at MIT where our goal was to understand how far into the future you could predict the advancement of science, based on current knowledge. It was set up in a pseudo-"think tank" format, with short presentations on topics about the future—where the next big advances would come from, the extent to which government funding accelerates progress, and the like. The topic would then be thrown open to the audience for participation, with a moderator posing questions.

The participants were academics, scientists and philosophers, representatives of R&D operations in high-tech and medical companies, a few government scientists and a handful of senatorial and congressional aides, as well as representatives of the White House OTA (Office of Technology Assessment).

After much bandying about, the answer we came up with was that about half of advances occurring within five years into the future can be predicted based on current knowledge and the use of the scientific method, namely, making a hypothesis and experimenting to test it. However, the other half can't be predicted because they come about through "eureka" discoveries. In other words, Turton's method.

It turns out that at least two of the companies at the MIT conference had professional science fiction writers on staff, in an effort to see if any fictional "lateral thinking" could be translated into real science. This wasn't as fanciful as it seems, because science fiction writers form conceptual science totally independent of existing methodology—a fertile ground for "eureka" developments.

The development of television is just one example of the "eureka" phenomenon at work. Television was developed before World War II, and it was obvious that at some point it would become widely utilized in the public sector. But at the time, television relied on vacuum tubes, which were not only expensive but unreliable. It wasn't until the advent of transistors—a "eureka" development—that television became a mass-market product.

So despite his drawbacks as a companion I learned something very valuable from Turton: that imagination and quantum thinking are the essence of major advances in science. This one insight was to have a major impact on my career.

I have no idea what happened to Turton after that.

Chapter Six

Airplanes

When I first applied to Nottingham University in 1954, I was intrigued with the University brochure. It showed three planes flying in formation, with the University as a backdrop. To me the planes seemed to be Spitfires, the famous World War II fighters the British Royal Air Force flew. The caption of the photograph was: "Members of the University Air Squadron in Formation over University Park."

Coming from a family of limited means, I had never seen the inside of an aircraft. Actually *flying* a plane was inconceivable. Instead I assumed that the planes in the photograph must have had University students as passengers. I never imagined that by joining the Air Squadron, I myself could really *fly* one of these planes.

But in my freshman year at University, I became friends with a guy named Mike Watts. Not only was he actually a member of the Air Squadron, but he told me that as a part-time member of the Royal Air Force Volunteer Reserve (RAFVR) he got to travel abroad and he was actually paid for the hours he spent flying. On top of that, he told me confidentially, "It's a great way to meet girls"—something else that was outside my own experience!

There were two "flights" in the Air Squadron. The "Technical Flight" consisted of people who learned about planes but didn't actually fly. I wasn't much interested in that. I wanted to be in the "Flying Flight," where students were actually trained to fly planes *alone!*

What Mike didn't tell me was that getting in was no cakewalk. It involved a technical and physical exam, and it was pretty selective. Fortunately I scrabbled through, and by my second year in college, I was spending one or two afternoons a week, depending on the weather, learning about planes and flying.

My first trip outside of England came about during summer break that year, when our squadron was carted off to the British colony of Malta. At the time the British Royal Air Force maintained a base at Luqua Air Base outside Valetta. As an officer cadet in the RAFVR, I lived quite the life. I had my own "batman" (a servant), ate and drank in the officers' mess, and had the use of the officers' beach. The local girls used to line up hoping that an officer or officer cadet would invite them onto the officers' beach. Nirvana!

This was everything Mike Watts had promised it would be. Not only was I flying—ecstasy!—but I had my pick of eager Maltese girls. Unfortunately, I didn't realize at the time that the local mores treated dating one of them as tantamount to being engaged. For some weeks after I returned to England I dodged letters, phone calls, and telegrams from Maltese parents and families, demanding that I marry girls that by British standards would have hardly been considered acquaintances.

Learning to fly planes in the Air Force is a bit different from private flying, mostly because you have to learn all kinds of stunts that private piloting doesn't require. They taught us how to stall and spin the planes, as well as how to carry out precision maneuvers like stall turns, slow rolls and barrel rolls.

Cadets typically had to spend nine hours of flight time with instructors before they could go solo, and in this respect I was totally average. Today, several hundred flying hours later, the concept of being alone in an airplane is still mind boggling to me.

While good three-dimensional co-ordination is the only skill you really need to be a pilot, these kinds of stunts required a strong stomach, as well. About half of the original recruits dropped out during the first year because the stunt flying put their guts through the spin cycle.

One poor fellow, who'd had many hours of solo glider flying, just couldn't get used to the G forces involved in the stunts, nor the noise or distinctive kerosene odor of fuel in powered planes. He'd go up for twenty minutes, land, throw out his full barf bag, and then get right back in the plane and try all over again. After about ten hours of logged time in the air, he gave up.

The plane we learned on was called a Chipmunk. It was a single engine plane with a fixed undercarriage, and it could do all kinds of aerobatics. The trainer would sit in the back, and I'd sit in front.

By today's standards the Chipmunk was very rudimentary. For instance, planes today have electric starters. Chipmunks started with an explosive cartridge that turned the engine. Chipmunks were tail-draggers;

that is, they had two wheels under the wings, and then a small wheel under the tail. That meant you had to land nose high; you couldn't see the ground, and that was a real problem. Today's typical trainer, a Cessna 150, has "tricycle gear"; that is, instead of the tail wheel, there is a wheel up front near the propeller, so that the undercarriage looks like a tricycle. And Chipmunks had low wings, or wings beneath the windows, whereas Cessnas have wings above the windows. As a result the vision is much better in Cessnas than on the Chipmunks.

For all of its primitive nature, the Chipmunk was just a terrific little plane. I loved it. It was very strong, reliable and almost unbreakable—a valuable quality considering what the cadets put it through!

While my first solo flight was blessedly uneventful, some of my classmates tested just how forgiving the Chipmunk really was. One of my classmates was so relieved to return to *terra firma* after his first solo flight that on touchdown, he stood on the brakes. The plane responded by burying its nose and propeller in the ground and standing virtually on end! Fortunately only his pride was hurt.

Another student took off on his first solo, only to have the wind pick up while he made a circuit of the field. In principle this wasn't a problem, because we weren't landing on a paved runway but instead a grass field. That meant we could shift landing direction so we'd be landing into the wind. However, the changing winds meant that this student was trying to land with a crosswind, and that's an interesting experience in a tail-dragger. With a high-wing plane, like a Cessna, it's much easier. You tilt the wings and the rudder so that you land on one of the three wheels. You couldn't dip a wing in a tail-dragger because the wings were too close to the ground and had the radio antenna attached underneath them. Instead, you had to point the nose partially into the wind and fly straight down the runway, "kicking" the plane straight at the last minute to avoid shearing off the wheels. If you kicked too early, you'd be blown off the runway. It just wasn't easy.

This particular student tried and tried to land for an entire hour. Finally, he was talked down by people shouting at him from the control tower. He climbed out of the plane white as a sheet!

Another student came in too low on a final approach, and his landing gear got entangled in telephone wires. He radioed the control tower for help, while the plane dangled upside down from the wires.

Of course, there were more serious incidents as well, although fortunately none that I knew about involved any fatalities. One mishap occurred during a practice maneuver involving low flying. I think we were taught to fly low in order to avoid radar. What makes it difficult is that it requires quick reactions and a real appreciation of physics. For instance, when you are flying close to the ground over hilly terrain, you have to take into account that a plane tends to continue flying in a straight line after the "stick" is hauled back, because of momentum.

Once an instructor was demonstrating this effect to one of my classmates by flying toward a hill and hauling back on the stick at the last possible moment. It turns out he left it a microsecond too late, and ended up in a high-speed stall a few feet above the ground. BOOM! The plane hit the turf and disintegrated. Like something out of an old silent comedy, the student and the instructor were left sitting fore and aft with only a cockpit around them. The rest of the shredded plane decorated the surrounding landscape.

The two most harrowing flying experiences I ever had occurred during my first year in the cockpit, after I'd logged about twenty-five hours in the Chipmunk. In fact they happened one right after another, in the same disastrous week.

The entire squadron was at summer camp, flying out of Doncaster in Northern England. I was flying the Chipmunk solo on a routine training run, at about 6,000 feet over the countryside. Suddenly I heard a distinctive blurp-blurp-blurp as the engine started to misfire.

Now keep in mind that most pilots *never* have an engine fail. That it might happen to me as a rookie pilot was almost inconceivable. But there was no mistaking what was going on.

As a pilot, you're trained to do two things in the event of an engine failure. Well, three, if you include praying like hell. First, you look for evidence that the engine is on fire, either because the engine temperature gauge is through the roof or there's smoke in the cockpit. If so, you dive quickly to extinguish the fire, and you prepare to bail out. If there is no fire, you notify the control tower of the problem and you start looking for a field to land in.

I immediately radioed the tower. "Victor November"—that was my military piloting identification code—"Ten miles south of field, with engine failure." It was macho not to show any emotion.

Just as calmly, the tower replied, "Roger, Victor November. We have you on radar. Keep us informed."

So I turned my attention to looking for a field to land in. In Northern England, those are in plentiful supply. There are really only two criteria you look for: the field has to be flat, and it has to be sheep-free. If it is too narrow to land normally, you are supposed to try to land so the cockpit stays intact but the wings break off. On top of that, you have to land into the wind to help slow you down quickly. There was no way to tell the wind direction in the air, and instead we were told to look for waves on ponds for some hint about where the wind was coming from.

I was descending quickly. Suddenly the engine cut out entirely. I was going to have to land pretty soon. With the altimeter edging closer to a thousand feet, I prepared to do the last thing I'd been trained to do: cut the ignition (to prevent a fire) and jettison the canopy. Just as I was about to switch off the engine, it partially restarted. Brrr POP POP POP. It was heavily misfiring, and I couldn't maintain altitude, but it gave me a hairsbreadth of a chance to land the plane on our normal landing strip.

"Victor November, I'm five miles out," I radioed the tower. "I have engine restart but I can't maintain altitude."

"Come in direct approach North-South, Victor November. Will hold all traffic."

I sputtered toward the field. I could see the Canberra Bombers circling above me, waiting to see what would happen. On either side of the runway was a parade of fire engines and ambulances. I hoped against hope that I wouldn't need them. At least there wasn't a hearse.

There was a hedgerow at the end of the runway closest to me. I cleared it by about ten feet. The engine still sputtering, I managed a normal landing.

When I jumped out of the plane, I didn't get quite the reception I'd expected. My squadron leader immediately berated me, "You should have landed in a field, Walton! You could have wrecked the plane and killed yourself trying to get back here." Perhaps he could see the dismay on my face, because his voice softened. "Of course—we're very pleased you got back."

That single incident would have been enough excitement for my entire flying career. But as Fate would have it, I had another scare just a few days later.

In the Royal Air Force, national examiners would randomly test cadet pilots to judge their level of training. In my squadron of twenty pilots, I was the one who was tapped that very week. I was absolutely petrified. After all, this was a wing commander, and it wasn't just my skill as a pilot that was at stake; I was representing the entire squadron. But in fact it was soon clear that the reason this particular guy had been chosen to do these random tests was that he had the ability to put cadet pilots at ease, a real back-slapping regular guy.

We got into the plane, me in the front seat of the Chipmunk, and him in the back. After we took off, we communicated by radio. He casually asked, "I know you've only had twenty to thirty hours. Any acrobatics?"

"I can do loops, wingovers, stall turns," I replied. All pretty basic stuff. I demonstrated the loops and wingovers with no difficulty. Then he wanted me to do a stall turn. A stall turn is basically like flying on the inside of a paper clip. You do a very steep dive and then a climb almost vertically. When you are pointing straight up and you still have some speed, you kick in full rudder. That turns the plane around sharply so that it is pointing straight at the ground. You head down and pull out into some other maneuver. The idea is to get the plane as perpendicular as possible as you go upwards, and to finish straight and level at some reasonable altitude.

I did this, and he responded, "Not bad. But you should have taken it more perpendicular before kicking on the rudder. Watch this."

He started the steep climb, but then he made a serious mistake: he waited too long to kick on the rudder. We didn't have enough speed to make the turn downwards. It left us in the most dangerous position there is: a hammerhead stall. The plane hangs in mid-air pointing straight up. Then it plummets backwards like a rock. What makes this so dangerous is that like most planes, Chipmunks aren't engineered to fly backwards; their tendency is to break apart. And needless to say, that's *really* bad news.

As we began to fall backwards, he said, "Oh, s—t. Hang on!"

We both hung onto the stick to keep all the flying surfaces flat, so the rudder and aelerons couldn't be torn off. After a few excruciating seconds, the plane flopped over into a vertical screaming dive, heading straight down. Every loose paper in the cabin flew around as we rocketed toward the ground. But at least we were flying forwards instead of backwards, and that gave him the chance to get control of the plane.

We leveled out at a couple thousand feet. When my heart finished its return trip from my throat to my chest and he regained the capacity of coherent speech, he said, "All right. Perhaps your stall turn was better than mine after all."

Clearing his throat, he added, "If you say nothing, I'll say nothing."

Needless to say, my report was spotless.

Chapter Seven

The Drain in Brains Flies Mainly West in Planes

Performing well at University was the only way to escape the shackles of the class system in England. That was common knowledge to my friends and me at Nottingham. It put tremendous pressure on us to do well on the two major exams that went into the bachelor's degree. One took place at the end of our first year, and the other at the end of the third.

The competition was excruciating, far different from college life in the U.S. In England in the 1950's only one or two students out of a hundred went to a university at all. A slightly greater number went to lesser technical and arts colleges, while the vast majority of teenagers finished their education at age fifteen or sixteen. At University, if you majored in math, you had only a ten-to-twenty-percent chance of graduating.

I decided not to major in math. Instead I focused on my longtime favourite, chemistry.

Getting a bachelor's degree in England at that time took three years. There were two or three one-hour lectures every day, in classes of about a hundred students each. There were twenty or thirty chemistry majors. We had to take three different kinds of chemistry—inorganic, organic, and physical, which was basically chemical physics. We had to pass a technical proficiency exam, which required that we translate published chemistry papers that were written in German. And on top of our class work and exams, we had a chemistry lab or two every week.

During one of these labs, one of my classmates spilled burning benzene on his trouser leg. His lab partner quickly grabbed the CO_2 fire extinguisher and opened fire on the burning trouser leg. For a minute or two all we could see was a fog of CO_2 foam and vapour. Finally the benzene boy emerged, with one trouser leg normal length and the other shrunk above his thigh, the polyester fabric reacting badly to the CO_2. Our relief at his well-being was quickly overcome by the absurdity of his appearance. We all dissolved in laughter.

Many of my classmates during the first year at University seemed to play the game "I'm smart enough to get through this exam with very little studying." In fact, they weren't. And failing first year exams was the death knell of their education, since without passing those exams they couldn't get a scholarship for second year, and without scholarships they couldn't afford to come back.

During that first year I quickly realized that classes were a complete waste of time. Our teachers were awful. Being British, neither my classmates nor I would risk showing our ignorance by asking questions, even though we had no idea what was going on. Physical chemistry, quantum mechanics, statistical thermodynamics—after attending a few lectures that were simply incomprehensible, it dawned on me that a far better use of my time would be to teach myself the material from the assigned textbooks. The strategy paid off. I easily passed the first year exams, and by the time I'd reached my third year, I scored double my nearest competitor in a couple of classes.

Although I could take shortcuts with classes, that wasn't the case with the labs. Our general physics lab was a source of particular frustration for everybody. The labs were time-consuming, you didn't learn much from them, and the equipment often didn't work properly. What we usually wound up doing was to make the measurements in the lab and then finish the calculations and theoretical work at home. We'd hand in our write-ups to the professor. They didn't get graded. Instead, we had to complete a minimum number of satisfactory experiments in order to pass the class.

My frustration with the physics labs led me to make one of the most ridiculous mistakes of my life.

I cheated.

Halfway through my second semester at Nottingham, I was working on a physics lab that had me flummoxed. It involved three different types of measurements that had to be made and fitted to theory. After I went through the procedure it was clear to me that either the equipment was giving faulty readings or I had misread the difficult-to-obtain numbers, since my results didn't fit the theory we'd been given in any way.

Several hours of calculation at home left me with what I viewed as two choices. I could either do the lengthy experiment again, or work backwards to find reasonable numbers to fit the theory, a technique that we called "dry-labbing." Another word for dry-labbing is "cheating."

It was late at night, and I didn't relish the idea of performing the lengthy experiment again. So I dry-labbed it. The problem was that in this particular case, it involved fudging numbers that were reasonably involved, and they required a three-dimensional iterative calculus that was well outside my experience.

After working all night on the project, I eventually came to what seemed reasonable conclusions. I handed in my lab book the following day.

A week later, the professor overseeing the course called me into his office. I knew immediately that I was in trouble.

"Walton," he said, "I would like to talk to you about your experiment last week."

My face reddened.

"Your results are very interesting," he continued, "but you should know that one of the measurements you reported is outside the range of the measuring instrument. In fact, no one has been able to get this experiment to work all year."

I was trapped. I had visions of failing the course. Losing my scholarship. Hanging myself from the nearest tree.

"But—"

But, I thought. *But what? You can always go back to Cadbury's?*

The professor continued, "As far as I can see, you have invented a new form of math in obtaining the solution. Please explain to me how you did this. I want to incorporate your method into my research!"

I have never been more relieved. I spent the next half hour explaining what I'd done, to his obvious appreciation and interest.

I didn't, however, get credit for the experiment.

I *did* pass the class. And I never, ever cheated again.

xxxxxx

As we approached the third year exams, the stress for many of my classmates became intolerable. We believed our futures hinged on our performance on that exam. The type of degree we got—"ordinary" being the lowest, and then on up through honors III, IIb, IIa, and I—would determine our opportunities. Envisioning the abyss of broken dreams and a life doomed to the lower rungs of the class system, some students killed themselves.

One of my drinking buddies, Bernard, had been the top student in his year one exams, but felt in his third year that he was losing his grip. He got drunk and drowned in a local canal. The president of the student philosophical society shot himself in his dorm room right before finals. One of my closest friends described in great detail how he would commit suicide if he failed. He didn't. Two years later, I was best man at his wedding.

With the third year exams looming closer, I was fairly confident I would get an honors degree. The question instead was whether I'd score high enough to pursue my goal of getting a Ph.D. and ultimately becoming an academic. This would require at least a second class honors degree. While it was a goal I eagerly sought, the alternative for me wasn't suicide! Instead, I thought I'd teach chemistry, or fly from carriers in the Fleet Air Arm. I liked the uniform.

As it turns out, I scored fifth on the exams, meriting an upper second class honor degree. That was enough to get me into the Ph.D. program.

I'd be lying if I said getting my doctorate was all work and no play. Playing bridge, flying, and my musical endeavours took up a fair amount of time. On top of that, I married Jasmin and we had baby Kimm. Nonetheless, I got my Ph.D. in 1960, after two and a half years of research.

Now I had to find a way to my career goal, which was to become a Professor. In England, "Professor" is a title that is only accorded to the head of a department. To get there, you have to start as an Assistant Lecturer, and then become a Lecturer (which would be considered an Assistant Professor in America).

The problem was that the only way to get one of these coveted entry-level positions was for a present lecturer to die, and to find yourself in the lucky position of being a close favourite of the professor who chose the next lecturer.

Those were lousy odds in my book. Instead it seemed a far more fruitful path would be to follow the example of my Ph.D. supervisor, Geoff Parfitt, and head for the U.S.A. We all knew that the streets of America were paved with gold. I intended to go and dig some of it up, and bring it back to the U.K. after a couple of years.

As a fallback, I interviewed with the main British employer of graduate chemists at that time, Imperial Chemical Industries, or I.C.I. as it was known. I went through a very degrading day of interviews. Every time I mentioned that I wanted to pursue a Ph.D., an interviewer would tell me I was wasting my time and that industrial training at I.C.I. was a far better preparation for life. This flew in the face of what I could see with my own eyes. All of the organic chemists I observed at I.C.I. were doing the menial work of lab techs.

Some time later, when I was at Indiana University, Professor Walter More told me that in his view, I.C.I. propelled more British Ph.D. chemists to emigrate to the U.S. than any other single factor. It certainly had that effect on me!

In the early 1960's, the U.S. policy of allowing unlimited immigration of British citizens led to a situation described by the phrase, "The drain in brains flies mainly west in planes."

In my case, it was a boat. But coming to the U.S. wasn't *that* easy. To go into academia I needed a post-doctoral sponsor in the U.S. I turned to Geoff Parfitt and got to know everybody I could through him. I pursued Victor La Mer at Columbia University and Frank Gucker at Indiana University, both of whom were prominent physical chemists. I knew La Mer's work very well, and Gucker supposedly had the world's first ultrasonic interferometer, an instrument that intrigued me. I didn't hear from La Mer. Gucker invited me to join him at Indiana, and I did.

Columbia University was a truly world-class institution, and had La Mer made me an offer, I probably would have taken it. Living in New York City would have had an incalculable effect on my family, and the Type-A nature of the city is not one I would have enjoyed. Indiana University, by contrast, was not nearly as prestigious. But it was also in the very livable, tranquil college town of Bloomington. Fortunately, since I'd only received an offer from Indiana, my choice had been made for me.

It was several years before I had the chance to reflect on the merest quirks that weigh so heavily on our lives, including the misdelivery of even a single letter.

I gave a symposium lecture in California in the late 1960's, and La Mer was in the audience. Afterwards, he came up to congratulate me on my performance. Then he added, "By the way—why didn't you ever respond to my post-doctoral offer?"

Chapter Eight

Westward Ho!

I arrived in the United States in 1960 with $10 in my pocket, thirteen pieces of luggage—and more significantly, a wife and a baby daughter.

I had met my wife Jasmin at Nottingham when I was in my third year and she was in her second year. We met in a very conventional way, at a university dance. As was typical for her, she was surrounded by a group of men. I doubt I would have had the nerve to speak to her had we not been introduced by a mutual friend. I was lucky enough to dance with her, and her conversation kept me laughing for hours. She was so different from the shy, unattractive, blue-stockinged girls who seemed to populate the university at that time.

She was in great demand on campus, dating only the wealthiest and most prominent students and townies; in fact, when we met, she was engaged to a rich naval cadet who lived in town. (I am not sure how I ousted him from her affections; it must have been persistence.) She was exotically beautiful, the product of a colorful family tree that reputedly included an Indian maharajah. She was intelligent, witty, but most alluringly of all, she was charismatic. One day she would be the life of the party, and the next, curiously aloof.

Jasmin had done things that could most charitably be described as quixotic. Perhaps her defining moment occurred when she was a first year student at Nottingham. It was customary to get the entire year's scholarship money at the very beginning of the school year. In Jasmin's case, this amounted to three hundred some odd pounds (which today would be about ten thousand American dollars). Having come from an achingly poor family, she had never had any kind of money at her disposal, let alone three hundred pounds.

She promptly went out and spent the entire sum on a spectacular designer dress, leaving her with no money for food, rent, or anything else.

But it was quite a dress.

We married in 1958 and became parents in 1959. It wasn't long before I realized that Jasmin's moodiness was more than just a reflection of a mercurial personality. It was the harbinger of mental illness, which was to lead her to suicide some years later.

Looming tragedy was the furthest thing from my mind when we arrived in the States. I had anticipated only two things—large cars and the American flag—and they were both greatly in evidence.

My Anglo-centrism was quickly challenged in several respects. One had to do with, of all things, building materials. English houses are mainly brick or stone, while their American counterparts are mostly wooden. I couldn't understand why people as wealthy as Americans would use an unstable material like wood, when surely they could afford the more durable brick. I quickly learned that wood was a much better insulator. In fact, the only reason English homes weren't wooden was that England had been virtually denuded of timber during the days of wooden ships and wood-burning fireplaces.

We arrived in Bloomington on the Monon Railroad. By our British standards, our apartment in married student housing was very luxurious. It was on the top floor of a brand-new three-story building, with two bedrooms, a living room and dining nook off the kitchen, and all the modern conveniences.

The friendliness of Americans was immediately borne out in all of the people who offered to help us move in—not that we had anything to move!

I quickly found that Americans were as ignorant of the British as we were of them. People asked us all kinds of rather unusual questions. Was rice the main staple food in England? Did British people have television? Where *was* England, anyway?

The most curious of my early encounters in Bloomington was with a local shopkeeper.

"You're a foreigner, aren't you? You must be at the University."

"Yes. We just arrived."

"What do you do?"

"I'm a scientist."

"Scientist? We don't need scientists. Everything is in the Bible, including Sputnik . . ."

A conversation like this would have been unimaginable in England. It made me realize that I wasn't too far from Scopes Monkey Trial country.

Very shortly after we arrived in Bloomington, we took our first plunge into Americana: we bought a big American car. It was a 1954 Ford Custom. It was enormous. The British cars of my University days could easily have fit in its trunk. Driving on the "wrong" side of the road in this boat wasn't nearly as difficult to get used to as the car's automatic transmission. Automatics were virtually unheard-of in England at that time, and I'd never driven one. Sure enough, on one of my first forays in the Ford, I unthinkingly depressed a non-existent clutch. Instead I was pressing on the brake with my left foot, while I kept my right on the accelerator. The car screeched to a halt and I nearly went through the windscreen.

One day I parked near a traffic sign that said "No parking on the pavement." It took a traffic ticket to teach me that the pavement in England is the sidewalk in the U.S., and the roadway in England is the pavement in the U.S.

Clearly my international driving license didn't mean I could drive like a native American. The license—which I'd received in London—meant that I could drive temporarily anywhere in the world. In due course, I went to the local Motor Vehicle Bureau to ask about applying for an American driver's license. The woman at the bureau asked, "How did you get here?"

I replied, truthfully, "I drove."

"How could you have done *that,*" she asked, "when you don't have a license?"

"But I *do* have a license. I can drive on any continent." I proudly produced my international license.

She examined it for a moment. "We don't accept these here," she huffed.

I took the license back, leafed through it, and pointed out to her that it said I was entitled to drive in the United States.

"But not," she replied triumphantly, "in Indiana." I had to be driven home by a "qualified" driver.

Over the next few weeks, I studied hard for my American driving test. I straightened out the pedal situation, and I learned the vocabulary. I drove to the test center, where I waited with a bunch of spotty teenagers to take my test.

When the test examiner greeted me he asked, "What kind of license do you have?"

Although I had an American learner's permit, and in spite of my earlier experience with the woman at the bureau, I produced my international license. The examiner took one look at it and said, "You don't need a test. Come over here and I'll issue you an Indiana license."

I wasn't the only British expat to face difficulties negotiating the American roads. One of my English friends was so used to driving in the U.K. that for the first few days, he looked for traffic lights on street corners instead of overhead. He was color blind anyway, so the location of the lights might not have made much of a difference. However he quickly learned to refocus his attention after he zoomed unthinkingly through a red light in his convertible—and a startled cop fired a warning pistol shot over his head!

One day my Ford was in the shop and I had to bum a ride to work from a French colleague's wife. Collette was probably the worst driver I have ever known, before or since, and so I knew that the two-mile ride to work would be an adventure.

In fact she drove what must have been the only stick shift car on campus, and even though she had some difficulty with it, the short trip was initially uneventful. We successfully dropped off her daughter at kindergarten on the way. As she bundled the little girl into school, I started to think maybe we were in the clear after all.

My complacency was soon shattered. When she got back into the car, she rammed it into reverse. Without looking, she stepped on the gas, and backed straight into the passenger-side door of a relatively new Ford, owned by a Chinese student who was dropping off his daughter at the same kindergarten. A furious Collette leapt out of the car to confront him. Since neither of them spoke good English, the exchange was rather comical. I will never forget his response to her demand: "Why did you illegally park your car behind me?"

"Madam," he responded politely, "do you crash into every car that is illegally parked?"

With her driving habits, Collette's car was clearly not long for this world. And in fact a few months later she demolished it. She and her husband were on vacation in Florida. While foolhardily allowing her to drive, he told her to turn left, without adding the key words, "At the next corner." She immediately turned left, taking out a fence and hedgerow in the process, before finally dropping the transmission in the middle of a field.

Chapter Nine

"God Is My Collaborator"

Life at Indiana University differed markedly from my expectations. I had been warned in London that the pace and pressure of work in the States was intense. I anticipated that I would be burning a great deal of midnight oil, proving to my new American boss what a great young creative scientist he had hired.

In fact, if America moved at a breakneck pace, it was a fact well-hidden from sleepy Bloomington, Indiana. I was greatly amused to find that it was possible to major in Sports. Sports! Students were actually able to get academic credit for playing golf. This was not just an ocean but a universe away from my own experience scrambling with every ounce of intellectual energy to stay at University at all.

My enthusiasm for working at the University evaporated almost immediately. I had come over on the promise that I'd be working on an ultrasonic interferometer, my excitement sparked by the fact that it was the only one in the world. In essence, an ultrasonic interferometer passes ultrasonic sound waves through a solution at different frequencies, temperatures and pressures. With this machine we could perform all kinds of experiments on the physical properties of solutions, testing hypotheses that were previously only possible to analyze theoretically.

However, what I actually got to work on was a microcalorimeter, which was a lot more basic. All it could do was compare the absorption of small amounts of heat for salt solutions and non-salt solutions. To add insult to injury, the microcalorimeter I got to use was a decrepit one which I was told to rebuild. Even if it was fully functional, it was only capable of measuring specific heat to a few parts per million, to fit a theory which everyone knew would work.

Fortunately, within six months of arriving at Indiana, I got the chance to teach. It was only Freshman Chemistry, but I was now called an "Instructor"—a bump up in prestige, at least in my mind—and it came with an increased stipend.

I was gripped initially with a fear of public speaking. My audience, fifty or so nursing students who clearly had no interest in chemistry, didn't make it any easier. I developed a series of pre-lecture rituals to calm my nerves, which included a bathroom break. One day in the winter of 1961, a particularly cold day, I was completing my bathroom ritual. When I tried to pull up my zipper, it broke off in my hand.

I was in a total panic! I had only been teaching for a couple of months. I did not feel sufficiently comfortable to finesse the situation. Instead I decided to lecture in my buttoned-up overcoat. I'm not even sure my class noticed, and if they did, they didn't betray their impression. They must have attributed it to an eccentric English behavior.

During another semester, when I was teaching Freshman Chemistry to a new crop of a hundred students, I had my first exposure to cheating. Bald-faced, serious cheating. I always thought that if I ever caught anybody cheating, they'd be mortified. I was in for quite a surprise.

The way that I tested my chemistry students was through a series of True/False exams. After each exam, I'd hand back the question books to each student and go over the answers.

After I handed back the first exam, one of the students cornered me and said, "Look. You graded five of my answers wrong."

I checked his book against my grading sheet, and sure enough, five of the ones that had been marked wrong were clearly correct. I adjusted his score, but not without misgivings. It seemed extremely puzzling to me that while a few students had pointed out that I'd marked one of their answers incorrectly, no one else had come anywhere close to finding *five* mistakes.

When it came time for the next exam, I xeroxed this particular student's answers before I handed back the question books. Sure enough, after the explanation session, he was back up in my face. "Look!" he said. "This time you made *seven* mistakes on my exam!"

I took his test book from him, assuring him I'd look into it. Sure enough, when I checked his original exam with the one he'd handed me after class, seven of the answers were different.

I notified my supervising professor. The student was given an "F" in the class and kicked out of school. But that's not the end of the story. He was reinstated after his rabbi successfully argued that the student's family had put intolerable pressure on him to become a doctor, and he needed an A in chemistry to get into med school. His F in the course remained. I have no idea if he eventually got into medical school. I certainly hope not.

I wish I could say that he was the only cheater I ever caught. But in fact there were a lot of them. Mostly they were premed students, under terrific grade pressure. And in general they came clean when they were confronted with the evidence. The two most difficult cases I ever had were students who stridently claimed they were entitled to cheat. It was their destiny to succeed because God had told them so.

I wonder, then, why God didn't tell them the right answers.

xxxxx

I wasn't the first British post-doc at Indiana University. In fact, when I arrived, I was surprised to find that at departmental parties Americans were singing English drinking songs they'd learned from my predecessors years before. I was greatly amused to hear Americans singing lewd English words, with only the vaguest idea of what they meant.

While I was at Indiana, I received a letter from Professor Eley, the head of the Chemistry Department at Nottingham, asking if I would look after a Ph.D. student of his who was going to be working for a chemistry professor at Indiana University. This particular student was the mellifluously named Ron Snart. Although he had been at Nottingham at the same time as me, my only recollection of him was that he was a nerd.

He turned out to be far more interesting than I had expected.

Waiting at the tiny Bloomington Airport, I expected to meet Ron and his pregnant wife. Ron emerged from the puddle-jumper by himself, explaining that he'd left his wife in England until their baby was born. I assumed that he did this in order to save money; in England, the National Health Care System would pay for the birth, while the cost in the U.S. was an unknown.

It turns out that Ron was very unpopular with his in-laws and undoubtedly wasn't welcome there. This was borne out when he made a very expensive phone call to England at Christmas to talk to his wife. His father-in-law, failing perhaps deliberately to recognize Ron's voice, hung up on him.

A divorce followed shortly after.

Ron was quirky in other ways. He was seriously color blind, and unusually so in that his color-blindness encompassed not just red and green but blue, as well. This gave him a dangerous penchant for running red lights.

His color-blindness was not a desirable characteristic for an aspiring chemist, for whom color recognition is vitally important. Ron dismissed his color-blindness by attributing it to his lack of experience with colors, presumably in contrast with people who make an effort to *learn* colors at an early age.

At the University Ron worked in a lab just down the hall from mine. True to my promise to Professor Eley, I dropped in to see how Ron was doing. He was conducting a DNA experiment which involved precise measures of changes in spectroscopy for a solution in a test tube. Partway through the reaction in the test tube, he dropped the tube. Instead of starting again—standard practice in chemistry—he calmly mopped up the solution from the floor with a dirty rag, squeezed it into another tube, and continued with the experiment as though nothing had happened.

It wasn't really any of my business, of course, but my sense of scientific ethics prodded me to ask, "Ron, what on Earth are you doing?"

He responded matter-of-factly, "I don't really need to do this experiment at all, because God has already told me the answer."

As it turns out, further questioning revealed that Ron claimed to have undergone a religious revelation while getting into bed one night (he did not mention the circumstances). Divine intervention aside, I found it hard to believe that work so unbelievably sloppy would escape the eagle eye of Ron's supervisor, Professor Moore, one of the world's leading research chemists.

God obviously intervened in Ron's career at least once more, because the next time I came across him was to read his name as the author of an article in the prestigious "Transactions of the Faraday Society," where only the most highly-regarded biophysical and chemical articles were published. Part of me was wounded that such an obvious nutcase could succeed. What made things worse was that Ron had written the article about the DNA experiment that I'd caught him mopping up in the lab. Hoping to find that the article wasn't very good, I took it to a colleague with expertise in the subject of Snart's article. To my great and everlasting surprise, he opined that it was an interesting and professional piece of work.

Perhaps God deserved credit as co-author.

Another infamous British post-doc resident in the Chemistry Department was a guy who'd left years before I arrived, but his legend lived on. I'll call him Sandy.

Sandy faced a problem that all of us at Indiana tussled with every day: parking. The Indiana University parking authority seemingly issued a hundred parking permits for the fifty or so parking spaces outside the Chemistry Building. While this encouraged all of us to get to the lab very early so that we could park, Sandy refused to do this. He liked to work late and sleep late. When he arrived at the lab all of the parking spaces were taken, so he routinely parked illegally outside the front door of the building.

He amassed a mountain of parking tickets and warnings. Finally one day his customary ticket included a threat that his car would be towed and confiscated if he parked illegally again.

He could have parked elsewhere or come in earlier the following day. But he didn't. He came in late, parked in his customary illegal spot, and recruited his friends to maintain constant surveillance for the campus police. The local cop eventually showed up, noted the presence of Sandy's car and rushed off to get a tow truck. In those days, the cop was the one who had to hook up and remove offending cars, all by himself.

When the cop returned with the tow truck, Sandy's friends took note and immediately alerted Sandy. He realized that quick action was necessary if he was going to save his car. Since he was on the third floor of the Chemistry Building, he divined that his quickest means of action would be gravity-driven. He grabbed a bucket from the lab, filled it with water, and promptly poured it out the lab window and onto the cop's head as he was hooking up Sandy's car to the tow truck.

The cop, furious and drenched, abandoned Sandy's car and rushed into the Chemistry Building to find the perpetrator. As he entered via the front staircase, Sandy ducked out the back, ran around to the front of the building, got into his car and drove off.

Without any hard evidence that Sandy was the water bandit, he wasn't arrested or even discovered. And as further evidence that crime does in fact occasionally pay, the story goes that he left the University before the authorities were able to deduct the parking tickets from his paycheck.

XXXXXX

For the first several months I taught at Indiana I was still struggling with the microcalorimeter I was supposed to be rebuilding. I finally gave up in disgust, and started redesigning the underlying electronics of the system. This was a snap. When I presented it to the man who was supervising my post-doctoral research, the improbably-named Professor Gucker, he said,

"It looks like it might work. But I'd like you to prove the theory using the original system!"

So back to work I went, with only a little more success than I'd had before. I wound up working on it for the entire fourteen months that I was at Indiana. I eventually cobbled together data that was about the most accurate in the world, but it wasn't significantly better than what other people using this godforsaken microcalorimeter had come up with before. And it certainly wasn't sufficient to prove the theory. I subsequently met the previous post-doc who'd worked on the microcalorimeter, a fellow Brit. He told me it was the most depressing project he'd ever worked on. So at least I wasn't alone.

Although my work in the lab was not going well, both Jasmin and I had come to enjoy living in the States and I thought we should try it out for another year or two. Of course after the year or two went by and we still wanted to live in the States, I rationalized my extended stay by saying that I would go back to England if the Queen or the Prime Minister asked me to.

So far that hasn't happened.

Although Professor Gucker wanted me to stay and work for another year on the microcalorimeter—which I increasingly viewed as the instrument of my torture—my thoughts and aspirations lay elsewhere. With my immigrant visa in hand, I was free to work anywhere I wanted. I quickly went about scouring ads and making contacts.

I had noticed several publications from Case Institute of Technology that related to my work at Nottingham, and they seemed to lend themselves to a much more thorough mathematical analysis. I sent a note with my ideas to Louis Klein, who was then the Dean of Science at Case Institute and the author of these publications. In retrospect, it was a ballsy thing to do. While more senior and recognized scientists write to authors suggesting new approaches all the time, I was far from senior and Klein could easily have shrugged me off. But he didn't. He invited me to lecture at Case. A few months later he called to offer me an Assistant Professorship of Chemistry, which came with the princely salary of $9,500 a year.

A difference in American and English terminology almost made me turn down the offer. In England, universities are considered far more prestigious than technical institutes, which led me to seriously consider a job at a truly backwater little American university. Fortunately for me, I explained my choice to a few American friends at Indiana, who said, "Are you *nuts?* What about M.I.T. or Cal Tech? You don't think they're 'worthy'?" They brought me to my senses before any damage was done, and I was on my way to Case.

Chapter Ten

Case the Pressure Cooker

As a student I always believed that you chose your career based on how you measured your own success. Businessmen measure their success in money. Doctors seek immortality. Psychologists want to solve their own problems. The coin of the realm in academia is intellectual achievement, and that's why I chose it.

In 1961, I was offered the position of assistant professor in the Chemistry Department at the Case Institute of Technology, in Cleveland, Ohio. I borrowed fifty dollars from my friend Louis Gordon so that I could move our household to Cleveland. I felt embarrassed about being too poor to finance the move myself. "Oh, no, you're lucky," Lou assured me. "The more money you make, the more money you owe."

I found that homily completely untrue in later life. I also quickly found that many aspects of academia weren't as I dreamed they'd be.

I expected that a university would be an intellectually stimulating environment, free of the politics and pressure of the business world. "Where else could we indulge our hobby and be paid for it?" mused my department chairman at Case Institute, Eric Baer. He was right. Research was just plain *fun*.

But academia has pressures all its own, as I discovered almost immediately.

When I arrived at Case Institute, I found that there were several hundred applicants for each faculty opening in chemistry. While I felt lucky to snag a position as assistant professor, I quickly learned that only two of the six assistant professors would ever make tenure as associate professors. These odds became far worse when "merger fever" hit Case; different departments combined, and Case Institute itself merged with

Western Reserve University in 1966, to become Case Western Reserve University. At that point half of the Chemistry Department was either fired or drifted away to other departments or universities.

And then there was money. For my first several years at Case, nearly all of my graduate students went to jobs making more money to start than I was earning after years of teaching.

While Case Institute gave me the glorious opportunity to dream up ideas and turn them to reality, I landed—thud!—in the reality of the American academic system and the three things that were expected of me: publications, publications, and more publications.

Here's why publishing is the lifeblood of academia. Research publications are read by scientists all over the world. As a result, the quality of a university and its faculty are judged by their research accomplishments as reflected in publications. The fact is, the world has little or no basis for judging the teaching accomplishments of faculty members. Consequently, universities rank in the ratings race based on publications and *not* on teaching quality, and students suffer as a result.

As a new professor, I found that academic publishing involved a problem I'd seen described in a P.G. Wodehouse novel. Wodehouse's hero, Bertie Wooster, once commented that getting started as a portrait painter was extremely difficult because nobody would hire you to paint their portrait unless you'd painted other portraits. But you couldn't get your first portrait commission without a portfolio. Likewise at Case Institute, I found that you needed research grants to conduct the research to produce the publications— but you needed publications to get the grants to do the research. It took me some time to learn a trick that academics still use today: do the research first to see if the idea works, and then write a grant proposal for it. With the grant money, you do the research for the next grant proposal you intend to write.

The grant proposals were typically submitted to the NIH (National Institutes of Health), NSF (National Science Foundation), and occasionally to industry. Obtaining grant money wasn't made any easier by the fact that I seriously questioned the vision of some of the grant reviewers. I remember one reviewer in particular opining, "It is impossible to get synthetic polypeptides to form crystals," and rejecting my grant proposal as a result. Convinced that he was wrong, I filched money from another grant to support a grad student on this research (a fairly common practice in academia). Sure enough, over the next few years, my clever graduate students were able to produce dozens and dozens of such crystals.

But academia isn't all research and grant writing. The activity that made my colleagues and me the most anxious, at least as new professors, was presenting research ideas and data at scientific meetings. While it's the goal of every professor to gird his graduate students to face a grilling at these meetings, there's no way to completely prepare anyone for the horrors of public speaking.

When I think back to some of the comments I heard lobbed at unfortunate scientist-presenters, I am most grateful that I wasn't the recipient. I have actually seen young professors faint from anxiety in the middle of their lectures. As a graduate student at Nottingham, I remember listening to a young scientist present his theoretical work based on the theories of Dr. Coulson, the great mathematician and chemist. His task wasn't made any easier by the fact that Dr. Coulson himself was in the audience. I hardly understood a word of the scientist's presentation, and believed I was the only one who couldn't figure out what he was saying.

I was wrong.

At the end of the lecture, Coulson stood up and, in front of a hushed crowd, said, "Tell me again, young man, *exactly* what you have done with my theory."

Of course, scientific conferences aren't a steady forum for humiliation. There are light moments as well. One I particularly remember involved two brilliant academics of my early years—Professors Mel Glimcher of Harvard and Aaron Posner of Cornell. They had opposing theories on how bone is formed, a delightful battle of ideas fueled by the fact that at that time, few of the mechanisms of biological chemistry were actually known.

Both Glimcher and Posner were present at a Gordon Conference in New Hampshire concerning bone formation. Glimcher, one of the most brilliant speakers of the era, made a truly spectacular presentation on the first day of the conference. While conference speeches rarely merit more than polite applause, Glimcher finished his lecture to a rousing ovation from the audience—by everyone except Posner.

When the applause died down and the audience members sat down, Posner clambered to his feet.

"Give me an aspirin, Mel," he said. "You give me a headache!"

The audience collapsed in laughter.

XXXXXX

As a new professor, it quickly became obvious to me that I wasn't interested in spending my career expanding on ideas I'd already researched. I found it much more exciting to take the lessons learned in one area and apply them to something else entirely. This had the incidental and wonderful benefit of making me an instant world expert on new areas of research.

For instance, my earliest research involved the question of how crystals formed from groups of atoms or molecules, and the mathematical equations that might describe these processes. After listening to one of my lectures on the subject, Professor Ralph Beebe of Amherst College commented, "You know, Alan, the same principles you talked about might explain what's going on with the way calcium phosphate crystals form in bone. Or cholesterol crystals in gall stones."

This suggestion led me to a whole host of new questions. How does bone protein, collagen, affect crystallization? Can you calculate protein structures by studying the coding sequence of their amino acids? What goes wrong with protein structures when people get sick and age? And on and on. Research was a constantly unfolding puzzle; as soon as you solved one problem, it illuminated other, more sophisticated problems to attack. It was endlessly fascinating.

Forming crystals from liquids and understanding everything about their formation is called "nucleation," and I quickly developed an expertise in it. Now the interesting thing with this phenomenon—the formation of crystals—is that even though it's not the kind of thing anybody pays attention to in day-to-day life, it's integral to all kinds of products. It's even in your body. "Good" nucleation helps make body parts that you want to be strong, like teeth and bones. "Bad" nucleation can be a real pain—literally. It can create gall stones and kidney stones.

In industry nucleation phenomena are very important indeed, because if you don't pay attention to how and when crystals form, you won't create what you think you're creating. For instance, if you want to make glass windows, you start with melted silica. But if you allow nucleation and crystallization to occur, you've got ceramics. There's nothing wrong with ceramics, but you wouldn't want them in the windshield of your car.

Another example is oil. When oil is pumped from the ground, it often contains carbon particles. If those particles are allowed to aggregate or clump, they could easily clog the world's major pipelines. Oil companies solve this problem by introducing minute amounts of oil-soluble polymers, which are components of plastic that block the nucleation, to stop the oil from clumping, and save the pipelines from clogging.

Plastics are an excellent example of nucleation and crystallization. Preventing or inducing nucleation completely changes the nature of the plastics that are created. Without nucleation, you have plastics that are pliable and extrudable, perfect for carpet fibers. Induce nucleation, and you've got plastics that are tough and strong, ideal for car parts.

So sometimes you want to promote nucleation, and sometimes you want to prevent it. And some substances are very difficult to crystallize, no matter how much you want them to. Among those are organic chemicals formed from carbon, nitrogen, hydrogen and the like. One apocryphal story has it that at the beginning of the twentieth century the famous German chemist Karl Fischer was able to crystallize many compounds by shaking his beard into organic solutions. Yuck! It turns out that the impurities from his beard—bits of last night's dinner, perhaps—acted as nucleating agents.

Another interesting feature of crystallization is that once you've crystallized a chemical you can always crystallize more, anywhere you want, by using some very minute amount of the original crystals to nucleate a new crop. That's good if you want more crystallization, but very, very bad if you don't. But whether you want it or not, once you start the nucleation process, it's almost impossible to stop it because minute crystals float through the air and cause a chain reaction of crystallization.

If that's not the stuff of science fiction, I don't know what is. In fact, during World War I, nucleation almost became a weapon of war. Here's what happened. Nitroglycerin, so important as a compound of high explosives, was being produced in glass form by the Germans. A British scientist proposed that the British government consider bombing German nitroglycerin factories with nucleating agents that would make the nitroglycerin unusable. It was theoretically a great idea, but remember what I said about the difficulty of stopping crystallization once it's started? Well, officials felt that producing the nucleating product in England might have disabled the *British* nitroglycerin factories. So they never pursued it.

With my expertise in nucleation and crystallization, I was able to help solve all kinds of problems, from medical puzzles to all kinds of industrial issues involving bread dough, lighter fluid, ice cream—you name it.

xxxxxx

Thank God for the intellectual triumphs at Case. They fed my desperate need to succeed. It was not, however, mere ambition that led me to work fourteen hour days, seven days a week. Spending virtually every waking moment working saved me from facing a home life that was ever more unbearable. My wife Jasmin became increasingly mentally unstable. Her descent into madness seemed both inescapable and unavoidable. As her outbursts became more frequent and her behavior more irrational, I became even more of a workaholic.

After a few years of working at a breakneck pace, my nerves were shot. I couldn't drink caffeine. Television—or any mindless diversion—was to me what food is to an anorexic. I couldn't stand it. My anxiety finally reached such a level in 1966 that my physician prescribed a two week vacation in the woods of Canada, camping and fishing.

I took along a couple of graduate students as well as my British friend Michael Henstock. It was the first time in my adult life that I'd had even one week off, let alone two. It worked miracles on my health, and on my fortunes, too. (As I write these words, I am ironically back in the woods—this time near the Canadian border, living in grander style, but still fishing and enjoying the woods, the sun and the lake. *Plus ça change, plus c'est le même chose.*)

On my return to work, I was promoted to Associate Professor, and with it I earned my tenure. This was particularly fortuitous timing. Shortly thereafter Case Institute merged with Western Reserve University, resulting in a massive bloodletting in the Chemistry Department. A few months the other direction, and I might well have been one of the unlucky ones.

Chapter Eleven

My Friend Michael

Few people—at least, few in my experience—have personalities so vivid that I can recall the very first time I met them.

My friend Michael Henstock is just such a person.

We first crossed paths as students at Nottingham University, where he was a year behind me. He was a friend of Mike Watts, my dormitory roommate. One day Michael showed up in our room to borrow Mike's lecture notes. In order to reach the notes, he had to shove aside beer bottles, unwashed socks, piles of books and other detritus of dorm life so that he could reach the desk wedged between our two beds.

To be hospitable, Mike and I offered him a beer.

"No, thank you," he responded, very properly, as he picked his way through the mess. "I only drink milk." *Milk!*

The rest of Michael's life has proven similarly eccentric.

Despite our differences, we became fast friends. Our first mutual interest was my girlfriend and later wife, Jasmin. Michael to this day tells my daughter Kimm that had things been slightly different, he would be her father. I'm not sure he realizes the strife I saved him by marrying Jasmin myself.

His interest in Jasmin aside, his main hobby was learning about, listening to, and lecturing about opera. Although the atmosphere at Nottingham was decidedly intellectual, it didn't encompass opera, and Michael was viewed as something of an odd egg as a result. What people didn't realize about him was that he has an outstanding sense of humour, and most importantly he is well able to laugh at himself. On top of that, he is a wonderful conversationalist because he has a towering intellect, an interest not just in opera but in widely-flung fields from world travel to adventure to history.

I used to think that Michael was a kind of magnet for amusing incidents. But I realize now that it is his singular personality that creates them. Fastidious in every way, Michael seems to hone in on everyday slights and inconveniences that most of us would shrug off. I inadvertently laid the groundwork for one of his memorable letter-writing campaigns by paving the way for him to visit the United States.

By way of background, in the summer of 1963, Michael had been an assistant lecturer at Nottingham in Metallurgy for a couple of years. I arranged for him to spend the summer at Case Western, working with my research group on a summer research fellowship. He planned to stay at our house during the week, and spend his weekends touring the U.S. on a travel-anywhere-in-the-USA-in-a-ten-week-period Greyhound bus ticket.

His visit with us was great fun. He introduced us to water pistol and food fights, and accused Jasmin and me of ganging up on him at Monopoly (we did; he would have won otherwise). Every Friday he would toddle off to the bus station in downtown Cleveland, get on a bus, take a sleeping pill, and wake up in some distant place. He saw much of America this way and enjoyed it immensely.

Because he was satisfied with these trips, they were not fertile grounds for correspondence. It was his attempt to recreate his journeys a few years later that provoked him to put pen to paper. By that time, he had married Anne, to whom he'd been engaged when he stayed with us. He wanted to show her America the same way he'd seen it: by Greyhound bus.

The trip did not go terribly well, a result Michael attributed to his conviction that the Greyhound bus company had changed the configuration of its seats. His frequent pronouncements on this apparently drove his new bride to distraction, but when Michael is fixated on something, there is no dissuading him.

When he and Anne got to Toronto, Michael discovered some older buses and set about very deliberately measuring the separation, inclination angle, and other dimensions of the seats, demonstrating to his satisfaction that the seats *had* indeed been changed for the worse.

Most people would be satisfied at this point, having been proven correct. Not Michael. He wrote to the president of Greyhound, sharing his data. The president responded that the seat configuration had indeed been changed, per the best consideration of Greyhound's engineers.

The correspondence did not stop there. A month or so later, Michael got a letter from Greyhound's chief engineer, indicating that they had reconsidered the situation and were changing the seats back to their old configuration.

So much for not fighting City Hall.

On the same trip, Michael and Anne arrived late in the evening in Chicago. They decided to stay overnight in a nearby hotel. Anyone who has ever taken a bus knows that the neighborhoods surrounding bus stations are not typically littered with Ritz-Carltons. Chicago was no different. They chose the best hotel of a bad lot, a rather seedy place with a sinister-looking apparent ex-convict acting as the desk clerk on duty.

In his best British accent, Michael asked politely to review a room. It passed his cursory inspection, but after he and Anne checked in he found that the toilet did not flush. He called the front desk. The clerk tersely advised him that there was no mechanic on duty, there were no other rooms available, and he could not get a refund. The deathly quiet of the hotel suggested that on the contrary it was, in fact, mostly empty, and the desk clerk's truculence in the face of this irked Michael. He told the clerk: "Either we get another room, or I shall be forced to call the British Consulate." This elicited some caustic remark from the clerk.

Michael thereafter set about trying to rouse somebody—anybody—at the British Consulate. I could have told him this wouldn't work, as I've tried it at British consulates around the world, and I've never found anybody at any hour, in any city. Stymied, Michael decided to call the police. It was apparently an off-night for crime, because after he described his plight, they told him to wait at the hotel door, and they would be there shortly.

Sure enough, within a few minutes a police car, lights flashing, pulled up in front of the hotel. Two burly cops jumped out. Michael, waiting at the curb, patiently explained to them his toilet predicament, and his brief tête-à-tête with the surly help.

The cops nodded and approached the bemused desk clerk. One of them grabbed him by the shirt, and said, "Listen, Pal, this man and his wife are visitors to this country. Either you find them another room or you give them their money back." The clerk suddenly discovered that there were, in fact, several vacant rooms, and Michael and Anne were soon happily ensconced in one of them.

When he got back to England, Michael wrote to Mayor Richard Daley of Chicago to thank him for the help of his police. Mayor Daley wrote back that he was delighted they had been of service, and added that he had passed Michael's letter to the Chief of Police.

A month or so later, Michael got a letter from the Chicago Police Chief, telling him that the cops in question had been identified and letters of commendation had been placed in their files.

His entire life has been littered with incidents like these. There was the time he pulled the emergency cord on a train in London—the emergency cord!—because of a smoker in a non-smoking train. Once he tussled with thugs in Sicily who were trying to mug him. And on one occasion he was flying in a commercial airliner that crash-landed in a field, and he settled with the airline for a new suit.

But it's not just the incidentals in his life that make Michael unique. He dedicated his working life to what he called "Garbology"—that is, recycling materials. And he got into it at a time when recycling was regarded as somewhat quaint and eccentric. He gained some renown in the field, travelling to remote parts of the world and publishing extensively. One of my favourites of his quotes is: "There is more nutrition in the box the corn flakes come in than in the corn flakes themselves."

His other—and primary—love in life was, and is, opera. For many, many years he pursued writing a biography of the turn-of-the-century opera singer Fernando de Lucia. Since virtually no one other than serious opera buffs had heard of de Lucia, it's not as though other aspiring biographers were nipping at his heels. He learned Italian in order to interview de Lucia's relatives, and made frequent trips to Italy to research his quarry. After he'd turned over every de Lucia stone for more than two decades, I was quite certain that the pursuit intrigued him more than the result, and that the de Lucia biography would never actually materialize.

But it did. The book *Fernando de Lucia* was published in 1995. It is widely recognized as the classic opera biography.

Chapter Twelve

I'll Go If You-go-slavia

After I graduated from Nottingham University in 1957, I spent the summer tootling around Europe with my friend Graham Neal, camping and generally seeing all we could, as cheaply as we could. We rode a Lambretta, a motor scooter with a 250 cc engine, which might have powered a very fine sewing machine but was taxed to the hilt by two riders and a load of camping gear.

We spent the first part of our trip in Western Europe, which wasn't all that foreign to us. But then we rode into Yugoslavia from Trieste, Italy, and entered an entirely different world. Newly divorced from the Soviet bloc, Yugoslavia at that time was hardly a tourist mecca. To make matters worse, there was a virtually unbroachable language barrier. Serbo-Croatian was totally incomprehensible to us. At the border we found that the only way to communicate with the border guards was in German—mine learned in school, theirs via the World War II occupation by Germany.

Once inside Yugoslavia we found to our dismay that camping was prohibited. This promised to put a serious strain on our virtually nonexistent finances. Between hotels and meals we anticipated a very expensive stay, and so we changed much of our travel money into Yugoslav dinars.

We spent our first night in Ljubliana, now the capital of Slovenia. It was small, poor, and drab. In fact, almost everything in Yugoslavia at that time seemed to be shrouded in gray. Dinner for two (perfectly edible, especially for someone like me whose only culinary experience was English cuisine), including wine, room for one night, and breakfast, cost us a grand total of about a dollar or two. Now then, things were looking considerably rosier! We had many more dinars than we would need to live on, and looked forward to spending more freely than we had in Western Europe.

We decided that a couple of days were all a fair investigation of Yugoslavia would require. Since there were no tourists there were virtually no shops selling anything we'd want to buy, so that didn't help deplete our dinar supply. The few commodities that did exist were strictly rationed.

The most memorable part of our visit was a trip down into the Postojna caves, reputedly the largest underground cave system in Europe. The Germans had apparently used part of the caves as an ammunition dump during the war, and the Yugoslav freedom fighters had snuck through the connecting cave system and blown up the dump. We were shown a huge blackened cavern as proof that the story was true. The caves were also useful for freedom fighters hiding from the Germans. All very intriguing stuff.

When the two days were up, we headed toward Austria, still quite flush with dinars.

At the border, a guard with a holstered pistol approached us. Since I was driving, he addressed me, in German, "How much Yugoslav money do you have?"

I answered truthfully. He immediately demanded, "Hand it over." I didn't get the impression that he was going to exchange it for me.

What cheek! I thought. I was both poor and resentful. I refused to hand over our money.

The guard immediately left and returned with a more senior guard, and we started our conversation over again. "Why are you trying to smuggle Yugoslav currency out of the country?"

"I am doing no such thing!" I protested. After a pause, the supervisor suggested rather forcefully, "You should spend the money at the local inn. Buy some slivovitz (a Yugoslav brand of plum brandy)."

Our knowledge of consumer pricing in Yugoslavia told us that we had enough money for about 120 bottles of the stuff. And anyway, we'd already tasted it, and weren't willing to drink any more. It tasted like a cross between lab alcohol and fuel oil, and perhaps it was.

Rather than insult the local brand, I protested, "Our scooter is already loaded down. The most we could carry with us would be one or two bottles." This clearly pushed the guards over the edge, because they lifted me off the scooter and frog marched me to the border post. Graham looked on, horrified and sure that he would never see me again.

They hauled me into a darkened office that looked like something out of a bad thirties spy movie, complete with a spotlight shining onto a chair in the middle of the room. The chief of the border police sat behind a dimly-lit desk in front of the chair. The only wall decoration I could make out was a portrait of the ubiquitous Yugoslav leader, Tito.

The guards deposited me on the spotlit chair. They relieved me of my passport and visa, and handed them to the police chief. My concern for my vacation money was by now overtaken by the growing perception that things were very serious.

The police chief, a heavy-set, barrel-chested man with the high cheek bones characteristic of Slavs, studied my passport in some detail. He muttered something in German about British students. Then the questioning started. "Why are you committing the serious crime against the People's Republic of Yugoslavia of smuggling Yugoslav currency out of the country?" My rudimentary German was barely adequate to argue that we were doing no such thing, and in any case, the police chief wasn't buying it.

He demanded my Yugoslav money, and by this time I was sufficiently frightened that I handed it over. The Chief grabbed it greedily, counted it and reached into a lower desk drawer. I fully expected that he was going to produce a submachine gun and execute me on the spot.

Instead, he produced a bundle of small denomination dinar notes, threw it at me, and yelled, "Get out!" I obeyed in a hurry.

Graham had been waiting by the bike, blissfully unaware of the drama inside. When he saw me race out of the guard post, yelling "We've got to get the hell out of here—get on the bike quick!" his reverie was shattered. We jumped on and sped over the border into more civilized Austria.

I subsequently learned that we'd done two things wrong. Well, three, if you count not handing over the dinars when the guard first asked for them. The first mistake was to lose our receipt for dinars when we entered the country; with this receipt, we would have been allowed to reconvert our exiting dinars into pounds. The second was that we didn't realize that dinars were barely convertible into hard, or western, currency, such that there was a ban on exporting dinars.

As we sped into Austria, this wasn't much on our minds. I felt fortunate just to be alive and not stuck in some Yugoslav prison. Based on the tourist hotels we'd stayed in, my mind boggled at what jail there might have been like.

But shortly our concern for more immediate needs washed away my relief. The bundle of dinars the police chief had thrown at me meant we had salvaged our meager savings. But when we took the dinars to the first Austrian bank we found, the banker laughed at us. "These are virtually worthless!"

Daunted, but determined to get *something* for our dinars, we found that the further we got from Yugoslavia, the better the exchange rate was. When we reached Holland, we were able to get about half of what we paid for them. Providentially, that was enough to get us a ferry ride back to England.

It would be fair to think that I'd learned my lesson when it came to Yugoslav currency conversion—or at the very least that the issue wasn't likely to come up again. But it did.

The genesis of my return visit came about in 1964. While I was Assistant Professor at Case, I hired a post-doc student, Helga Furedi, who was a staff scientist at the Rudjer Boskovic Institute in Zagreb, Yugoslavia (now Croatia). The Institute was very much like the National Institutes of Health in Bethesda, Maryland, or the Weizmann Institute in Israel. Fortunately for me Helga spoke good English, so my increasingly rusty German wasn't necessary. Working with Helga led to a fifteen-year on-and-off relationship with the Rudjer Boskovic Institute.

A year or two after Helga started working with me, a State Department official visited us and told us that because the U.S. had a large trade surplus with Yugoslavia, we could use part of that excess for a joint research project based in Yugoslavia. Neither Helga nor I was very enthusiastic. She thought Yugoslavia would be suspicious of a CIA connection. I wasn't too wild about facing the communist bureaucracy that had treated me with unique hospitality during my last visit.

The lure of research money eventually became so attractive that we set our concerns aside. In 1967, I became director for the joint project. The position involved writing the research plan and submitting the grant request, and then going over to analyze progress on the research once a year.

That summer, I visited the Rudjer Boskovic Institute for the first time. Jasmin and I made the trip into a bit of an adventure by picking up a Triumph sports car in England and taking part in an Alpine car rally through France, Germany, Austria and Switzerland on our way to Yugoslavia.

Once we got to Yugoslavia, we had to drive to the U.S. Embassy. The embassy was in Belgrade, now Serbia, while the work itself was in Zagreb. We got lost in Belgrade because, although we had a map, the street signs used the Cyrillic alphabet. To Westerners, it was (and is) incomprehensible.

The deal was that I would pick up my stipend at the Embassy, and that would cover our living expenses for a month in Zagreb. We were supposed to stay at the Institute's marine facility on the Adriatic coast, a primitive set of buildings whose main purpose was to monitor pollution emanating from rivers and other sources on the Adriatic coast and Italy. I don't remember exactly how much the stipend was, but even by U.S. standards it was generous, all in newly-printed 10,000 dinar notes.

My limited exposure to Yugoslav currency had taught me that we'd have a difficult time getting much out of the country, especially in denominations this large. So we happily determined that we would live as lavishly as possible and use up as much of the stipend as we could, rationalizing that we couldn't take much of it with us, anyway.

As it turns out, our efforts to spend money were stymied at every turn. For a start, contrary to the Spartan quarters we'd expected, we were given luxury accommodation courtesy of the Institute, so we didn't spend any money on rent. And on top of that, the Institute arranged invitations for me to give lectures all over the country, each of which involved a substantial honorarium. I wish I could say that my reputation merited same, but I suspect that my hosts saw me as a meal ticket to large American research grants and spoiled me accordingly.

On top of that, as had been true when I visited with Graham in 1957, there just wasn't a lot to buy. For the first time in my life, I was accumulating money much faster than I could spend it.

So my currency problems weren't resolving themselves. Our friends at the Institute were sympathetic—to the extent you can be sympathetic with anyone who has more money than they can spend—and tried to sell our dinars to visitors in exchange for hard cash. But there weren't enough visitors to make a dent in our stash. Of course, we could have given it away, but at the time I was still a poorly-paid junior professor and just couldn't bring myself to shovel it out.

Finally, Jasmin took matters into her own hands—or more precisely, her cups. She decided to smuggle the money out in her bra. On the one hand, my experience with the border police made me apprehensive. Visions of Yugoslav jail were running through my head. But on the other hand I thought her plan was nervy and exciting, so I went along with it.

Halfway through the summer we were due to visit our kids Kimm and Keir, who were staying with my parents in England. It seemed like a perfect time to execute the bra plot. We figured that the best chance to get the loot out was through the Zagreb Airport on our way to London, since it was the most direct route and time was at a premium.

As we proceeded through customs at Zagreb, I was predictably asked, "How much Yugoslav currency are you taking out of the country?" I had prepared myself for this question, and answered some nominal sum with an air of relative confidence. After all, I wasn't the one wearing the bra stuffed with bank notes.

When we arrived at Heathrow Airport in London, we hurried over to the Lloyd's Bank in the terminal to cash in our dinars. I asked the banker for the conversion rate, and he responded, "It depends on the denomination." One thousand dinar notes got a slightly higher rate than the five thousands, and he didn't even bother to quote us ten thousands.

I asked about the oversight. He paled and excused himself to see his manager. The manager came back, and told us rather haughtily, "I have never seen a ten thousand dinar note. But it would be convertible at the same rate as the five thousand."

I sheepishly responded that "It's not exactly an 'it' we're talking about." At this point Jasmin nonchalantly reached into her bra and produced about fifty newly-printed, albeit slightly wrinkled, ten thousand dinar notes.

The bank clerk was too stunned to speak for a moment. Then he sputtered, "Good God! Don't you know that removing this amount of money from Yugoslavia is highly illegal?"

I shrugged. After much muttering about flooding the market, and the fact that this amount of Yugoslav money would satisfy the conversion needs of several busloads of British tourists, he reluctantly handed over the sterling equivalent and we left, happy.

xxxxxx

My favourite experience in Yugoslavia occurred a few years later. In the early 70's, I was still working with the Rudjer Boskovic Institute. I was asked to give a series of three lectures sponsored by the Yugoslav National Academy of Science and the Croatian Chemical Society.

As usual, the Yugoslavs were outstanding hosts. I was chauffeured around in a Mercedes owned by the Institute. I was amused to find that my scientific colleagues, contrary to their normal loquaciousness, would not say a word in the car. The chauffeur, I gathered, worked for the national security service. While Tito's iron grip on the country had these kinds of unpleasant ramifications, everyone I knew agreed that when Tito died, the country would fragment into violently nationalistic subunits.

The first two lectures went smoothly. To celebrate, my hosts decided to take me out to lunch before my third and final lecture. In fact I wasn't at all prepared for the lecture, but figuring that the lunch was at noon and the lecture wasn't until four o'clock that afternoon, I was confident I'd have plenty of time to shuffle together a few slides.

As the lunch unfolded, it quickly became obvious that I'd have to scuttle those plans. Contrary to the quick sandwich I'd expected, there were about twenty guests—including the President of the Croatian Academy of Science, the President of the Rudjer Boskovic Institute, the President of the Croatian Chemical Society, and a bunch of other people I didn't want to offend by bailing out early. The meal itself stretched to seven courses, each accompanied by a different wine. I had no intention of drinking, conscious of my speaking engagement looming ever closer.

But with each course, my hosts would raise a toast to my health, and I in turn toasted *their* health. This endless round of toasting went on until 3:30, at which point I was feeling no pain. I was still sufficiently conscious to rationalize, though. I told myself that my condition wouldn't much matter because hardly anybody would actually show up for my lecture, I could throw the slides together in five minutes in the back of the Mercedes, and besides, I knew the stuff backwards.

I arrived at the hall to find it absolutely stuffed to the gills with scientists eager to hear what I had to say—that is, of course, if I could get my mouth to form words. In fact, I was very relaxed for the first fifteen minutes or so, totally unruffled by the fact that some of the slides were out of order or upside down. But as the effects of the wine began to wear off, panic set in. Perspiration started to trickle down my forehead.

As the slide problem grew more chaotic, I reached into my professorial bag of lecturing tricks and began to extemporize. While the slides were being constantly reshuffled, I told jokes. To my surprise and relief, the audience laughed in all the right places, after a momentary hiccup while they translated from English to Serbo-Croatian.

When the lecture ended I was a spent shell, exhausted and fearful that I'd been a failure. As one of the senior scientists from the Institute approached me, I expected—and deserved—a dressing-down. He held out his hand, smiled, and said, "You know, Alan, that was the best lecture I have ever heard you give." I was awarded the Rudjer Boskovic Medal (recognizing outstanding contributions to promoting Yugoslav science) for that performance. It still hangs on my office wall. I treasure it, although not for the reason most casual visitors expect!

xxxxxx

In the mid 70's, I was invited to an International Conference on something or other at the brand-new Hotel Croatia in Cavtat, just south of Dubrovnik. When I arrived, I was dazzled and believed that I had to put all my preconceived notions of Yugoslav hotels to pasture. Many of the places I'd stayed in were ones where I'd be reluctant to board my dog, what with their rudimentary rooms and stone-age plumbing. Clearly the Hotel Croatia was a radical departure from all of that. The main lobbies, the modern art, it was all stunningly beautiful.

But the plumbing was *still* lousy. My toilet flushed all night. The faucets dripped. The plug in the bathtub didn't work.

At breakfast after a harrowing night, I sat with a German scientist I'd never met before. By way of a polite opener, I asked, "How's the plumbing in your room?"

"Mein Gott!" he responded. "It is awful!" He raised a cautionary finger, and with a smile, went on, "But you haf to be prepared when you come to Yugoslavia." Whereupon he opened his jacket, and from an inside pocket he withdrew a beautiful miniature set of wrenches, screwdrivers, a soldering iron, a regular do-it-yourself plumbing kit.

When I returned to the Hotel Croatia a couple of years later with my new bride, E.J., I was hopeful that the plumbing problems had been corrected. And to some extent, they had. Instead, the air conditioning didn't work. I asked the hotel manager what the problem was and he responded, "Oh, the President of Peru is visiting President Tito in Belgrade, and they needed part of our compressor for their air conditioning."

Chapter Thirteen

Intrigue Among the Test Tubes

When I was a little boy, my vision of scientists involved men in white coats, standing in labs, surrounded by all kinds of beakers and test tubes filled with bubbling liquids. Within the first decade or so of my scientific career I found that science could, and did, involve a great deal more worldly intrigue.

In the mid 1960's, during the Krushchev era, the Cold War was in an extreme state of saber-rattling. It was very, very difficult for Westerners to travel to the Soviet Union and its satellites. Rather remarkably, tensions eased up enough at one point to allow for an international meeting of biophysicists and biological chemists in Moscow. I was dying to go, but I had not yet built up the academic stature to merit an invitation. Instead, I had to settle for savoring the many stories that emerged from the conference. I am convinced that the antics of the academics at that conference made the Soviet Union reconsider the desirability of normalized relations with the West.

My favorite story from the conference turned out to have nothing at all to do with science. It seems that conference attendees spent their days in session at the University of Moscow. At night, they were housed at the Hotel Ukrania if they were lucky, or student dorms at the University if they weren't so fortunate. The evenings were strictly monitored by Intourist guides. The principal activity, from what I heard, seemed to be drinking vodka in the local bars.

Everybody assumed that the KGB had bugged all of the hotel rooms and that they were eagerly monitoring conversations of the conference attendees—although after spending evenings drinking Russian vodka, I wonder how coherent those conversations could have been. Excess imbibement undoubtedly contributed to the experience of two U.S. scientists. They were sharing a room at the Ukrania, and got back to the hotel after a long night of drinking, determined to find the hidden microphone in their room.

Apparently the lodgings were quite ornate at the Ukrania. This made for a long and arduous search. They poked, prodded and pried at the chandelier, the imitation fireplace, the light fixtures, the walls behind portraits—no bugs. When they started lifting the carpets, they hit pay dirt. They found wires leading to a tube in the floor. Clearly this was some kind of bug! One of the scientists took out his pocket pen knife, and used it to sever the wires. The raw ends of the wires were stuffed into the tube in the floor. Satisfied that they had foiled the KGB, they headed for bed.

In the morning, they went downstairs for breakfast. The breakfast room was crowded, but it was immediately clear that something big was up because of the unusually loud buzz of conversation. They asked their friends at the table what all the talk was about and they were told: "Haven't you heard? When Professor Jones was in bed last night, the chandelier over his bed collapsed on top of him!"

It was many, many years before another major scientific symposium was held in Moscow.

Chapter Fourteen

The Tarnished Jewel in the Crown

One of the best aspects of being a scientist in academia is that as your reputation develops, universities all over the world invite you to visit and talk about the work you're doing. I saw it happen to colleagues of mine, and in the late 1960's and early 70's it started happening to me. I was invited to all kinds of places that I'd only dreamed about. Like India.

It was impossible to grow up in England and not succumb to the allure of India. Aside from its status as the jewel in the crown of the British Empire—which ended abruptly with India's independence, when I was twelve years old—books and movies about adventures involving the Khyber Pass and the Indian Mutiny gave India an exotic luster. Indian postage stamps, with their majestic elephants and imperious, jewel-laden maharajas, featured prominently in my stamp collection. My Indian connection grew more tangible when I married Jasmin, who was a quarter Indian and reputedly the great-granddaughter of an Indian maharajah, the Maharajah of Alwar. The story was that her grandfather was a Norwegian sea captain who ran off with a Maharajah's daughter—a dashing contrast to my own blacksmith-and-hotelier ancestors! (While the story of Jasmin's ancestry is very hard to verify, there's no question that her father, John Christensen, lived most of his life in India and acted as personal secretary to the Maharajah of Alwar, a position which is apparently often given to offspring.)

So when I was invited to visit Delhi, Madras, and Bangalore on a lecture tour in 1973, I looked forward to it with great anticipation.

My arrival in Bombay was a cold drink of water, but unfortunately not in the literal sense. The airport was extremely hot—a fact I should perhaps have anticipated for India in the middle of the summer. To my dismay the airport was considered cool because it was "air conditioned," which meant that there were ceiling fans, although with the heat of the air it was rather like mixing tar.

For some ungodly reason, planes arrive in and depart from India in the middle of the night. I flew in from Israel at 2 A.M. and expected to find a virtually shuttered Bombay airport. In fact it was astonishingly teeming with life. Rows and rows of benches, ostensibly for air travelers, were packed tight with sleeping families. I quickly learned that the airport was one of the few places where homeless people could find shelter in a place that was protected from both the weather (by those ceiling fans) and criminals. Benches overflowed with humanity. There wasn't a seat to be found for my two-hour wait for my flight to Delhi. So I grabbed my bags, stacked them up, and had myself a makeshift seat.

At the University of Delhi, I was told that my hosts had kindly arranged for me to stay in a bungalow on campus, complete with an in-room refrigerator and air-conditioning. The truth was a bit more rustic than the description. There was, in fact, an air conditioner in the window, but a cursory poke at it revealed that it probably hadn't worked since the Raj was dismantled. The attached bathroom had a rusty pipe protruding from the wall, with a temperature control that, at every setting, spat out water at one temperature: tepid.

During the day, the temperature made the room feel like a steam bath. At night, it dipped all the way down into the nineties. Sleep was an impossible dream. The only way to get any rest at all was to soak myself under the rusty shower, lie down, and close my eyes while the water evaporated. This way I could catnap for half an hour at a time.

Fortunately there were a few cool drinks in the bungalow's refrigerator. I had been warned to stick exclusively with bottled drinks in India, and that's what I did. Alcohol was virtually nonexistent because it was illegal to locals. I read that a local judge had actually proposed that drinking alcohol be considered a crime punishable by death! (Years later, India had a prime minister who announced that drinking urine was good for one's health, giving India a rather unusual set of drinking priorities.)

When it came to food, I wasn't as concerned with what was safe, because it was so hot that I had no incentive to eat anyway. On top of that, I had always hated Indian food. (The only way Jasmin introduced our children Kimm and Keir to Indian food was to cook it for them when I was away travelling.) So for my three days in Delhi, I was exhausted, hot, and starving.

I mustered the energy to lecture at the University on my second day in town. Afterwards, professors at the University took me out to lunch, despite my protestations that I really wasn't hungry. I had no idea how I was going to force down even one bite, but I felt compelled to go along, especially after they assured me that they were taking me to the best restaurant in Delhi.

We were seated at a big, round table. I again insisted that I didn't want anything substantial. Believing I was only being polite, my hosts ignored my protests and ordered me a whole chicken. A whole chicken! They each ordered one of their own as well. At this point I was really in a pickle. I knew that restaurant food was very expensive in India, and that this lunch was going to put a huge dent in my hosts' limited budget. But at the same time there was simply no way, in my weakened state, that I was going to be able to force down a chicken wing, let alone the whole bird.

I was still in a quandary when the chickens arrived. I took one look at the bird plopped down in front of me and thought I must be hallucinating. There could not have been more than three ounces of meat on the entire carcass. This poor mangy bird, with a crispy skin stretched taut over its bony frame, looked just like a chicken skeleton stuffed into a semi-inflated balloon. It had undoubtedly died of starvation. My concern immediately shifted from how I would eat a whole chicken, to exactly what I was supposed to eat at all. I glanced furtively at the other chickens on the table and realized they had met the same fate. I watched my hosts for a clue on how to proceed. They happily broke off bones and sucked out the marrow. I did the same thing, and scrabbled from the bones what tiny morsels of meat I could find.

Years later, Jack Koenig, a friend and colleague of mine from Case Western, visited India to attend a conference. As with all of my other friends, he'd heard my Indian chicken story and believed—as I would have if I'd heard it—that I'd been exaggerating. He, too, was taken to an expensive restaurant and treated to the house specialty, emaciated chicken dinner. He took one look at the pathetic pullet and exclaimed, "My God! It's Walton's chicken!"

xxxxxx

I went from Delhi to Bangalore, a charming colonial town in Southern India. I spent a delightful weekend there with one of India's most eminent scientists, Professor G.N. Ramachandran. He was partially responsible for discovering the structure of collagen (the main protein fiber found in tendons, bone and skin).

G.N. and his family—consisting of his son, a well-known physicist, and his daughter-in-law—took me out for a drive in the country and a picnic in the Nandi Hills above Bangalore. This was quite a treat, especially by Indian standards, since in the 1970's very few private individuals in India actually owned automobiles. It was an indication of G.N.'s revered status that he had a car, albeit a 1950's model which was considered plush by Indian standards.

The car broke down before we'd gone very far. Neither G.N. nor his son had any idea how to fix it. I reflected, to myself, that this was proof that the intellectual elite were better in theory than in practice. But then I couldn't figure out how to fix it, either. I poked around a bit. The engine turned over with no difficulty, and we had plenty of fuel. And unlike everything else in India, the engine wasn't overheated. I thought there might be an air bubble trapped in the fuel system, but without a tool kit this diagnosis was both unverifiable and unfixable. We needed help.

Because there were so few cars on the roads anywhere, I settled down for a very long wait. I had never heard about anyone dying while waiting by the roadside for a ride, but images of the emaciated chicken came to mind. G.N. and his family seemed unconcerned, and we sat down and picnicked by the side of the road. After about an hour, a taxi which must have predated World War II chugged its way up the hill toward us, hauling about a dozen passengers. We flagged down the driver, a very eager fellow only too willing to help. He tugged at a piece of plastic tubing leading from the carburetor, sucked gasoline through it with his mouth, reconnected it, and we were instantly in business. "Merely a vapor bubble," he proclaimed, and was on his way.

Despite the very edible picnic with the Ramachandrans, by the time I got to Madras, my last stop in India, I was distinctly unwell. I was overcome by "Delhi belly," sometimes gastroenteritis, sometimes dysentery. Again my accommodations were luxurious by local standards—a guest house in downtown Madras, with a bedroom and a cold-water bathroom. The room's most obvious feature was a large, open space in the wall where a window might once have been.

With the ninety degree heat coupled with ninety percent humidity, the room was a perfectly functioning sauna. Adding to my discomfort, the room was right next to the street and one floor above ground level, so there was the cacophony of what seemed like millions of chattering, banging, yelling people passing below my window. Even if I could somehow have forced my eyes closed in the heat, the noise would have kept me awake.

So by the time I lectured at Madras University for the first time, the scheduled "one hour plus questions" might as well have been an Olympic marathon. With visits to the men's room fore and aft, the lecture lasted a grand total of twenty-nine minutes.

The second day, when I was supposed to lecture again, followed another sleepless night. My green pallor must have made it obvious to my hosts that I was not in fighting form. That second lecture lasted twenty-four minutes, the shortest one-hour talk I have ever given.

I groaned back to my hotel room after the lecture, scheduled to spend two more days in Madras. I was certain that I would die before that and I was in such agony that it wasn't an unattractive outcome. By the evening I was in bed, delirious, oblivious even to the 4 A.M. call to prayer from the local mosques. (Although India is a Hindu country, with their Moslem brethren to the north in Pakistan regarded as mortal enemies, there were still many Moslems in the major cities at the time.)

At around 7 A.M., I was still lying motionless, face up, in bed. A big black bird flew into my room and hovered over the bed. In my mentally depleted state I hallucined a vulture come upon its soon-to-be-breakfast. In a fleeting moment of clarity I realized that if I didn't get down to breakfast and drink a gallon of tea, I was done for.

Somehow I staggered downstairs and gulped down all of the tea I could manage. It did make me feel marginally better. I crawled back to my room, only to find that both the big black bird and most of my belongings were gone. I had been robbed, and I doubted the big black bird was the culprit. Someone had apparently climbed through the window and made off with pocketfuls of cash and gifts.

The robbery was the last straw. I had to leave India immediately. The hotel called a cab for me. The driver pulled up in what looked like a World War I vintage taxi, and loaded my remaining belongings into its trunk. The driver then ritualistically kicked the front driver's side tire three times, and knelt down and prayed next to it. I watched curiously, but when we got underway I understood immediately his reliance on prayer. The four-way intersections had no traffic lights, no stop signs and no traffic police. It was every man for himself. I sat terrified, clinging to the seat, doubting three things at once: that I was physically sound enough to make it to the airport, that the rattling rustbucket of a taxi would hold together, and that regardless of the sanctity of either me or the taxi, we'd survive the killer-bumper-car traffic. Despite my atheism I began to wish I'd prayed with the driver!

When I mercifully made it to the airport, I took the first plane out of Madras. I vowed I would never return, which means, of course, that I ultimately *did* go back.

Chapter Fifteen

The Night I Didn't Spend in Taiwan

A few of my colleagues had told me that when they traveled to Eastern bloc countries, they were often briefed and debriefed by the CIA. One such colleague, who was due to visit Prague from Dubrovnik, found his trip being cancelled at the last moment by the Czech government. We had little doubt of the reason for that.

I didn't think that the CIA would ever take any notice of my foreign visits. But they did. In 1977, I was due to make a round-the-world trip heading westward to Taiwan, India (again!), and back through Europe to the States. Before I left, the "Company" visited me at Case Western. Now, I idolized James Bond as much as the next guy, but actually stepping into his well-polished shoes was something I greeted with apprehension. While Bond would have jetted in, snagged the goods while speaking perfect Mandarin Chinese (having taken a first in Oriental Languages at Oxford), and escaped with a beautiful Chinese girl on his arm and the bad guys firing at him and missing badly, in reality I was racked with nerves.

My assignment was fairly simple, and in fact didn't involve lapel cameras or, really, intrigue of any grand sort. I was to visit China and give a lecture, in English (thank God). While there, I was to learn what I could about what would now be known as chemical or biological warfare. The place where they wanted me to speak was the Institute for Biological Research in Shanghai, which was widely regarded as the most advanced in China. (Today, it is still very highly esteemed, and undoubtedly does have the chemical and germ warfare capabilities that the CIA suspected at the time.)

I was skeptical that the Institute would invite me to speak, but my visitors smiled and assured me that this would not be a problem. I questioned whether the Chinese would grant me a visa, and again, I was told this would be handled for me. My biggest question—how I would go about getting the information they wanted—was never made clear to me. Nonetheless, I was told that my background in both chemistry and molecular biology was "ideal" for the endeavor, and so I was set to go.

Sure enough, the Company quickly engineered the invitation, appropriate visas, interpreters and the like. Beyond that, I was on my own. I spent many sleepless nights envisioning life in a Chinese chain gang, shackled to a bunch of other unfortunate scientists, hacking away at plants in a rice paddy. The closer loomed my departure date, the more apprehensive I became.

To make matters worse, I had finally found the woman of my dreams, E.J., and had been married only five months. I didn't like the idea of taking off on a long trip and leaving my new bride behind. She couldn't help ease my fears because I was forbidden from telling her about the details of my mission. And even if I *had* been able to talk about it, I wouldn't have wanted to frighten her—or perhaps risk having her talk me out of going.

As luck would have it, the Chinese government suddenly, and for no apparent reason, cancelled my trip to Shanghai shortly before I was to visit. But my trip wound up being exciting for entirely different reasons.

At the time, I was a consultant to the White House, and as a result the State Department was picking up the expenses for my trip. While China had been the government's primary interest, they also thought it would be a good idea if I stopped off in Rhodesia and South Africa to see what progress they were making with the use of advanced technologies, like genetic engineering and germ warfare, while suffering economic sanctions from most of the world. They bought me a round-the-world ticket on Pan Am, and assured me that Pan Am and the government would make all necessary travel arrangements.

My first stop was to be Taipei, where I would lecture at the Taiwan National University. I left Cleveland a couple of days early, figuring that I'd lose a day crossing the international date line, and then I'd have a day to poke around before I had to lecture.

As luck would have it, the flight leg from Seattle to Taipei was interrupted due to weather. A typhoon over Guam forced us to land in Honolulu. This was a lovely surprise since Honolulu is one of my favourite places, and this was, after all, an all-expenses-paid stop. Pam Am put us up at a rather ritzy beachfront hotel for the night. (No doubt these kinds of extravagant delays helped to put Pan Am out of business.)

The one downside was that the delay in Honolulu was going to make me miss my speaking engagement in Taiwan. I sent a telegram to my host, the Dean of Sciences at the Taiwan National University, and received a very gracious wire back saying that he understood completely and would meet me at the airport anyway.

I finally arrived in Taiwan late Friday evening. At immigration, I presented my new American passport—as a White House consultant, I had been told that it would be a good idea if I became an American citizen! The immigration officer looked at it for a moment and demanded, "Where is your visa?"

"I was told I didn't need one," I replied. "I'm only going to be here for forty-eight hours."

"Who told you that?" the immigration officer shot back.

"Pan Am travel."

He shook his head furiously. "They never should have let you on the plane in Seattle without a visa. You must have a visa." I pulled out my British passport, thinking that maybe that would get me in without a visa. In fact it only confused them more, and the conversation was getting increasingly heated.

I realized that finger pointing wasn't going to get me anywhere, so I asked, politely, "Where can I get a visa?"

By this time, other immigration officials had joined the first guy, and they pointed out that it was late Friday night and the visa offices were closed. Instead, I had to get on the next plane to anywhere. Since I'd already missed my lecture, this wasn't a terrible alternative. I scanned the overhead flight board and saw that there was a flight due to take off within a half hour or so bound for the Philippines, and figured I'd hop on that.

But it wasn't so easy. The immigration police were huddled among themselves, chattering to each other and gesticulating in my direction. After a few minutes one of them came over and told me, in broken English, "We collect your bags and take you downtown under house arrest." Visions of dank rat-infested prison cells, knife-thin windows and tin cups floated in my mind.

Accompanied by the increasingly ominous immigration police, I went to baggage claim. In the distance, I could see a man holding up a sign reading "WALTON." I asked one of the guards if he would let this gentleman know what was going on. He agreed, and took off to speak with the sign fellow, whom I took to be my host. After listening to the guard for a moment, he waved happily to me and disappeared. 'He's not the one spending the night in prison,' I sniffed to myself, somewhat put off by his merry demeanor.

After picking up my bags, one of the immigration officers led me out to what I assumed would be a police car. Not so. He bundled me into a taxi and accompanied me to the Mandarin Hotel in downtown Taipei. I didn't check in personally; the immigration officer did whatever explaining there was to do about the "mystery guest." Instead, the officer escorted me to my room and locked me in, presumably keeping guard from next door. I had no idea what the following day would bring, nor whether in fact when—if ever—I would be allowed to leave Taiwan. Had it not been for this considerable dark cloud, it would have been an enjoyable stay, since the Mandarin is a very nice hotel, made even better by the fact that I wasn't paying for it.

Although I have slept fitfully all my life, on this one night—when I had more reason than ever to lie staring at the ceiling all night long—I slept quite well. The following morning, I was awakened by a phone call from my host at Taiwan University, saying that he and my guard were waiting downstairs for me at breakfast!

We had a very pleasant meal. The guard sat, stony-faced, as my host and I chatted about science, and how he might go about getting more of his students into U.S. universities—a recurring theme in my foreign travels. I successfully fought the temptation to suggest that giving up the practice of arresting visiting American academics would be a helpful start.

After breakfast, the guard escorted me back to the airport. To my surprise and delight, I was given my passport back, with the admonition: "You've never been in Taiwan."

My parents on their wedding day in 1927.

Mother and me during the War, when I was four.

Planning a squadron flight, Royal Air Force Volunteer Reserve in 1957. The plane is the sturdy and forgiving Chipmunk. I am on the far left.

The ten thousand dollar dress. Jasmin and me at a RAFVR Function in 1957. Jasmin is in the solid white dress, the third woman from the left in the front row. I am just behind her.

Elvis Walton

MOST people who were at the recent Air Squadron Dance or at Union Ball will already be familiar with the singing of Alan Walton, but it appears that his stylish renderings of the latest Rock 'n Roll numbers may soon be reaching a much wider audience, for as a result of his performance at the Air Squadron function, he has been offered an introduction to the well-known B.B.C. producer Wallace Eaton.

ALAN WALTON WITH A CHARACTERISTIC EXPRESSION.

What might have been. At the height of my music career, in 1958.

The plane that refused to come back to earth. Me standing in front of my Piper Colt, 1973.

With Ephraim Katzir, President of Israel, at Ayelet Hashachar in 1973.

Me with my beautiful E.J. and Miss Universe (Miss Sweden) in 1984. E.J. is the one without the sash and tiara.

Must… Catch… Fish…
White water rafting in Montana with my son Keir (far right) sitting next to me, and my friend Randal Charlton in front. This is the trip where Randal risked his life fishing through the rapids so he wouldn't be the only one without a fish, 1986.

At 14,500 feet on Mount Kilimanjaro with Bill Kokot (left) and Michael Irving (center). We look better than we felt, 1989.

Perhaps there is a God after all! Me having just survived my first free-fall sky dive, 1993.

Receiving a pretend honorary Ph.D. for "Biotechnology and Economics" at the 1996 National Conference on Biotech Ventures. The presenter is Ken Lee of Ernst & Young.

Preparing to marry my daughter Sherri—as the officiant, that is, 1997.

A non-life-threatening fishing expedition in Belize, 1999. Me with my son Keir (far right) and my son-in-law Paul (2nd from right), and our expert fishing guide sitting next to me.

#190 09-06-2011 12:09PM Item(s) checke

TITLE	BARCODE	DUE DAT
The biotechnologists and the evolution of biotech enterprises in the USA and Europe	39047078572423	09-27-1
Beneath this gruff exterior there beats a heart of plastic : my life in genetic engineering --and beyond	39047078254022	09-27-1

Chapter Sixteen

Back into the Belly of the Beast

After the excitement of my house arrest in Taiwan, I arrived in Singapore with mixed feelings. It was New Year's Eve 1977. 1978 was to be my first full year with my beautiful new wife E.J., and I missed her very badly. My trip had been arranged long before we got married in August of 1977, and I felt duty-bound to go even though I desperately wanted to be with her instead. But if I had to be away from her, Singapore was near the top of my list. It is a city I've always loved, partly because it's so modern and so clean, and partly out of British pride since Britain had had the opportunity to be the guardian of one of the world's great ports. And, of course, it was air-conditioned, a huge plus after the brain-melting heat of India four years earlier.

I wished there were some way to take the cool air, bottle it and carry it with me, because my next stop was India. My first visit to India had convinced me never, ever to set foot on the Asian subcontinent again. It had taken me several days in Japan and Hawaii to recover from that trip. But events conspired to change my mind.

I was invited to speak at an interesting International Congress of Biopolymers, convening in Madras, the day after New Year's in 1978. On top of that, I was asked to open a new extension of the Madras University in a town called Tiruchirapali, a hundred or so miles outside of Madras. Both were very attractive offers, if only they'd been somewhere else, *anywhere* else. There was also the matter of convenience; since I was travelling at State Department expense on a round-the-world ticket, I couldn't beg off due to the cost. But I was still loathe to accept, due to my horrid memories of India. I explained as gently as possible to my potential Indian hosts that I had been, euphemistically speaking, unwell on my last visit. I wasn't sure I wanted to risk the same fate. They reassured me that *this* visit would be very different, because my first trip was in June, before monsoon season, and "Most tourists get sick then"—a fact that had somehow been left out of my itinerary on my first visit. January, when I would be visiting this time, was immediately after the monsoons, when everything was washed clean. On top of that, they assured me that they would accommodate me in truly first class hotels, where "Westerners should be very comfortable."

"Anyway," they told me, "the train between Tiruchi and Madras is air-conditioned." I wasn't going to be suckered by *that*—it meant overhead fans which had the air-cooling effect of a passing moth fluttering its wings. But I found out that Indian Air traveled twice a week between the two cities, and that—on top of encouragement from the State Department—sealed the deal.

As luck would have it, my plane from Singapore arrived in Madras at the gate next to the Tiruchi (short for Tiruchirapali) departure gate. I knew this because of the cardboard sign, hand-lettered over the doorway. I approached a nearby official to see whether I could get on the plane. I was somewhat concerned by the fact that although a seat had been booked for me, I had never received a confirmation on this particular flight.

"Oh no, sir, it has been booked for months," he said, pointing outside to a twenty-seat twin-propeller plane.

"Well," I explained, "we sent you six telegrams, but you never responded."

"What is being your name, Sir?" he asked, running his finger down the list of anticipated passengers. "Oh, yes, Sir, I see there are six of you. Go right ahead."

xxxxxx

In Tiruchi, the fact that it was 'cool' season meant that my hotel room didn't get much over eighty degrees during the day. The room itself was cavernous and comfortable, and came complete with a friendly lizard climbing its walls. I had in fact found that lizards were a fairly standard fixture of hotels in India. I shied away from chasing them out for fear that they were regarded as good luck or perhaps a sacred reincarnation. I am not a particularly superstitious person but in India I felt I needed all the good fortune I could muster. In fact I had many discussions with Indian friends over the years about religion. I found it—and still find it—difficult to reconcile the abject poverty of Calcutta, Bombay, and many of the rural villages with the palliative that "things will be better in the next life."

I anticipated that Tiruchi itself would be a small town, largely because I'd never heard of it. But I quickly realized that there simply aren't small towns in India. After all, a small town in a country with 700 million people would be a fair-sized city in the U.S. At the opening celebration for the college I anticipated there would be forty or fifty diffident attendees. I would say a few inspiring words, they would applaud graciously, and that would be that.

Instead, I was led into a sizeable auditorium where the Vice Chancellor of the University introduced me to the crowd. I looked out over a sea of several hundred eager young faces. I launched into my fifteen minutes of prepared remarks about how education frees the mind, spirit and body from the most menial tasks, very appropriate in the circumstances, I thought. I hadn't poured my heart into my speech but nonetheless I was a bit put off by the sparse polite applause that followed my remarks.

The Vice Chancellor cleared his throat, and asked very tentatively, "We do not see many people from the U.S.A. or Europe. I wonder whether you would mind answering questions?" While I am very used to taking queries, what followed was totally unexpected. Students got up eagerly and asked things like "What is it like to have a washing machine?" "Why is the U.S. stealing all of our physicians?" "Don't you think we would be better off under communism?" I was quite taken aback but it was ultimately the most fascinating question-and-answer session I've ever had.

And, come to think of it, it is pretty wonderful having a washing machine.

After the formal Q & A finished, a few dozen students gathered around me to ask even more questions. Many of them wanted my help getting into U.S. universities, which at that time required sponsorship by a professor at an American school. Since I had never heard of Tiruchi, I was quite certain that they had never heard of Cleveland, and at that time—in the late 1970's, before the Rock and Roll Hall of Fame and the revivification of the Cleveland Indians led to an urban renaissance—that was something of a relief, because Cleveland was widely regarded in the U.S. as a real pit.

In fact, one of my biggest laugh-getting openers when I traveled was to say, "I am very pleased to be here today. In fact, I'm pleased to be anywhere other than Cleveland." The truth was very different. I very much enjoyed living there and found the suburbs as pleasant an environment as anywhere I've ever lived. But boosting Cleveland reached its most difficult point in the late 1960's, when its heavily polluted Cuyahoga River caught fire. Of course it was only an oil spill on the river that ignited, but the story caught the imagination of every newspaper editor in the world and it was reported with glee. Ten years later, the river had been brought back to life so spectacularly that coho salmon were being caught in it. Even then, I had a hard time persuading colleagues at the Harvard Medical School that you couldn't actually walk out onto the garbage floating in Cleveland's waters.

Back to Tiruchi and to the students eager to come to the U.S. One of them said to me, "Oh my goodness, Sir, you are coming from Cleveland, that is where the river is catching fire." Needless to say, he wasn't offered a position at Case Western.

After Tiruchi, I flew back to Madras for the Biopolymers Conference. Remembering my last trip to Madras, I returned with the greatest trepidation. But my fears were not realized. The hotel where the conference was held was delightful. And the food! It was worlds better, largely because now meat was plentiful—much to the relief, I'm sure, of the beleaguered Indian chickens. On my first visit, the fact that Hindus held cows sacred and Moslems considered pork filthy had conspired to make most of Southern India vegetarian, a diet contrary to my tastes. In the meantime they had discovered that lamb was an acceptable substitute for beef. And since there were no sacred sheep in India (and the idea of the lamb of God didn't faze them), all the hamburgers were actually lamburgers, and were quite tasty.

The conference itself was a great success, and I was well-received—or maybe I considered it a great success *because* I was well-received. And best of all, everything was air-conditioned, changing once and for all my opinion of India.

Chapter Seventeen

Straight Flush: Toilet Paper Around the World

When I started travelling extensively I developed a fascination with, of all things, toilet paper. One colleague told me that in post-World War II France, the shortage of toilet paper was so acute that they processed it from orange crates and other rough woods. As a result it was not unusual to find, embedded in the toilet paper, splinters up to six inches long!

I hadn't ever given toilet paper much thought until then. But I got to thinking that toilet paper might reflect a lot about local technology and social habits around the world, so I decided to start an international toilet paper collection of my own.

My travels in the mid 1970's gave me a perfect opportunity to develop my collection. In 1975, I traveled East from Cleveland, to New York, London, Munich, Kitzbuhl, Athens, Crete, Tel Aviv, Bombay, New Delhi, Madras, Singapore, Bangkok, Hong Kong, Tokyo, Honolulu, and back home. In 1976 and '77, I went westward to Seattle, Taipei, Singapore, Madras, Bombay, Salisbury, Nairobi, Johannesburg, Cape Town, Rio de Janeiro, New York, and back to Cleveland.

I found that there were indeed some textural differences from country to country, but not nearly what I'd expected. There were also variations at different locations within each country, with the best toilet paper, predictably enough, to be found at the finest hotels.

I was greatly embarrassed to find that by far the worst toilet paper in the world was in public toilets in my homeland, England. The texture of the paper is ill-suited to its purpose, something like a cross between wax paper and flimsy air-mail writing paper.

The best was in hotels in Hong Kong. Not only was the paper itself superior but it was presented very elegantly, in an exquisite outer wrapper. When I took a closer look at the wrapper I found that the paper had been manufactured in, of all places, England. I'd always heard that the Germans export their best wines. Apparently the British do the same with their toilet paper.

While texture was fairly uniform everywhere, what *did* differ immensely from country to country was the size of each sheet of toilet paper. The largest were Austrian, with Germany close behind (as it were). Apparently the purpose of the large sheets in Germany was to accommodate imprinted coats of arms—at least, that's what I found at the universities I visited. Examining these sheets, I wondered if the practice dated back to the Third Reich where, with a swastika on every sheet of toilet paper, there were always flags if needed.

The Austrian and German papers were more than twice the size of the smallest ones I found, which were in Israel. I assumed that this had something to do with the shortage of trees in Israel, and the country's need to preserve foreign currency by minimizing imports—of trees, toilet paper, or anything else.

Mind you, the very existence of toilet paper at all is no guarantee in some parts of the world. I found that in China and South America, newspapers were the standard *papier du derrière*. When my daughter Kimm and I took a train through the French Alps some years later, she visited the ladies room at a small café and in lieu of toilet paper, she found a telephone book with half of its pages torn out. I understand that the popularity of the Sears Catalogue in the U.S. in the early 1900's could be traced to the same utilitarianism.

I found that Indian public toilets had no paper at all, so either you brought your own or learned to live without.

When I returned to the States and shared my toilet paper collection with friends and colleagues, they volunteered their own toilet paper experiences. One friend told me about a trip he and his wife took to Russia in the 1970's. They were lucky enough to stay in a large hotel in the Ukraine, where the rooms actually had their own bathrooms. Alas, their bathroom had no toilet paper. They called the maid and asked for some by gesticulating creatively, since they didn't speak Russian and she spoke no English. Finally she nodded and pottered off down the hallway. They heard her footsteps as she entered various rooms along the corridor. After a few minutes she came back, triumphantly bearing a half-used roll of toilet paper. When my friend reached out for it, she held it back. She unrolled about twenty sheets from the roll, tore them off, handed them over, and headed back to replace the roll itself to its original room.

I have no idea whether these idiosyncrasies in toilet paper still exist—perhaps globalization has reached the commode. I suspect that the United States must produce much of the world's toilet paper, because it has such a huge wood pulp and paper processing industry. If this is, in fact, the case, I believe it is one of the most useful and least recognized weapons in thc American arsenal. We have seen countries hold out for years against weapons and trade embargoes, and even gas cut-offs, but a toilet paper embargo would surely bring any populace to its knees, so to speak, in short order.

Chapter Eighteen

Whispering Through Israel

My introduction to Israel, in 1973, did not get off to an auspicious start. I had been invited to the Israel Academy of Science in Jerusalem for a conference sponsored by the Rothschild Family of France. When I left Cleveland for New York, I had a slight case of laryngitis which got worse with every passing minute. By the time I boarded the El-Al flight to Tel Aviv, the laryngitis was severe and I was feverish. The flight was to last eleven hours, and I badly needed to rest and recover before I got to Tel Aviv.

I didn't get the opportunity. I was seated next to two elderly ladies, kvetching loudly to each other about the flight, and the food, and the service, and every half hour or so, poking me in the ribs and asking, "Are you all right?" On top of that was a gentleman seated nearby whose religious sect required that he lead prayers, in a very loud voice, every couple of hours, so as to ensure that no one be able to sleep on their way to the holy land. No wonder stewardesses regard the New-York-to-Tel-Aviv flight as the "hell flight."

By the time we landed in Tel Aviv I could no longer communicate verbally. My throat felt as though I had gargled with Drano. To make matters worse, my travel plans had been changed, and no word had gotten through to my Israeli hosts, the conference organizers. I had no idea of where I was supposed to go. Desperate for some rest, I headed for the nearest cheap hotel to sleep, regain my voice, and gather my wits.

I did find a reasonable-looking place nearby. Once I settled in, I realized I was famished. I noticed to my dismay that the hotel's dining room wasn't open. I asked in a whisper at the reception desk about where I ought to eat, and was told, "Oh, nothing is open today. It's Passover."

Famished, feverish and voiceless, I wandered around and found a shop or two open, apparently to sell essentials. Unfortunately, the two things I needed—food and toothpaste—were not considered essential. Starving and exhausted, I stumbled back to the hotel to sleep.

After napping for a few hours and regaining some of my strength, I again ventured out to forage for food. In the center of Tel Aviv I found a hamburger joint selling Wimpy Burgers. Food at last! But that was the good news. What I got was a dry, gray patty, sitting naked on unleavened bread. Undoubtedly it was the worst hamburger I ever ate. However, ordering a burger and a cup of tea with no voice at all was a real triumph—accomplished through whispering, croaking, writing and pointing—even if eating it wasn't.

The following morning, I was still stranded in Tel Aviv with no idea of where I was supposed to be, and no plan for tracking down the conference organizers. I remembered, however, that one of my colleagues at Case Western had given me the address of his relatives in Tel Aviv, encouraging me to call them if I got the chance. With a whisper of my voice returning, I went by cab to the address he'd given me. It was a row house rather like a smaller version of similar brownstones in London or New York. An elderly lady met me at the door. I took her to be my colleague's aunt. Since she spoke no English and I didn't speak Hebrew—and could only whisper in English—we tried to communicate in sign language, but to no avail. "I know your nephew" is not amenable to universal gestures. Fortunately, after a few fruitless minutes of this, a cousin showed up who spoke fluent English. Things looked up immediately. She called the physician who served the British and U.S. Embassies, and he prescribed antibiotics for me. She also—bless her—was able to track down the conference organizers, who'd apparently notified the police that I was a missing person, perhaps kidnapped by Arab terrorists. They promptly sent a car for me, and I was hauled off to the conference site in Jerusalem.

Jerusalem is a simply remarkable place, one of my favourite cities in the world. This was the first time I'd ever seen it. Every corner, every turn, every view left me awestruck. The old sector of the city, with its stunning Roman architecture. The Arab markets, where you can buy artifacts thousands of years old. The Via dela Rosa, where Christ was supposed to have hauled the cross. The twelve stations of the cross, each worshiped devoutly. The garden of Gethsemane, where there is an olive tree reputedly over 2,000 years old that Christ may well have seen. The Wailing Wall which, whatever faith you have (or lack), is extraordinary because of the overwhelming aura of devotion enveloping it. Thousands of notes are stuffed into cracks in the wall, requesting divine intervention. The atmosphere of reverence everywhere in the city is something I've never found anywhere else in the world, and it affected me greatly.

My driver took me directly to the conference. I was a little late for the opening cocktail party, but otherwise I hadn't missed anything. My laryngitis was on the mend, but I still sounded very hoarse. This was noted by a man I met, a lecturer from the Imperial College in England. "You could easily have cured your laryngitis by taking heavy doses of vitamin C," he assured me. I was already familiar with this idea, as was most of the world—it had been proposed a year or two earlier by the double Nobel Laureate chemist Linus Pauling. Since its introduction, the theory had fallen into disfavor in academic circles. I didn't know whether to take this fellow's advice seriously or not. I mumbled a noncommittal response, ultimately a good thing because I later learned that the gentleman giving me the advice was Linus Pauling's son.

Because I was still slightly feverish, I skipped the dinner that night. It was hosted by the mayor of Jerusalem, Teddy Kolik. The following morning I was wakened by a most unfamiliar sound. Although I had never actually heard the sound of cow bells, I assumed that must have been what I was hearing. But in the suburbs of Jerusalem?

I threw open my curtains and beheld a sight I will never forget. There was a shepherd tending a flock of sheep on the hillside which stretched right down to the hotel. The bells on the necks of the sheep furnished my serenade. The scene could have been lifted straight from biblical times, as there were no buildings, cables, or any other indicia of the twentieth century that I could see. I immediately felt uplifted and, in the way these things often work, also felt better physically than I had in several days.

At the conference itself, I was scheduled to give a lecture on my birthday, April 3. The conference organizers presented me with a bouquet of flowers as a birthday present. But the better news came later, when I received a medal for the most innovative piece of research presented at the conference.

I talked about the structure of proteins. There was, and is to this day, difficulty in explaining exactly why the proteins and molecules that make the body work are folded into their unique shapes. There had been many advanced mathematical theories, and I was able to show that simple models of proteins folded according to the rules of thermodynamics—a branch of mathematical chemistry that had not been used before. In my lab at Case, my grad students had proven this by working out the structure of simple biopolymers (which resemble proteins) using electron microscopes and diffraction.

What made my award for explaining this at the Jerusalem conference even more satisfying was that when I had sought a grant for the work the previous year, the reviewing scientists had turned down my request, saying, "Your experiment cannot work. It is impossible."

When we were not involved in scientific sessions at the conference, the Rothschilds saw to it that the thirty or so invited scientists were treated royally. We were whisked by luxury coach to the Sea of Galilee, Bethlehem, and the Dead Sea, as well as the old port city of Jaffa. Professors of archaeology and history from various universities gave us remarkable accounts of the historical treasures we were seeing.

The most memorable place we visited was Masada, where, early in the first millennium, Jewish zealots had held out for months against the Roman army. There are still massive round stones in the mountaintop fortress, which the Jews hurled down on the attacking army. After a long siege, the Romans built a ramp up to the highest part of the walls. The ramp was almost a mile long. When the Romans arrived, all of the survivors had committed suicide. The only person the Romans found alive was a small child cowering in the water cistern beneath the ramparts.

The fortress at Masada is in the desert overlooking the Dead Sea. Of course, we all had to go down and swim. The Dead Sea got its name from the fact that it is so salty that no animals or fish can live in it. In fact, the water has long since been completely saturated with salt, and there is a salt layer several feet deep at the bottom of the sea.

The surface of the sea is very shiny. And the salt makes it incredibly buoyant. You can easily sit on the surface of the water, with your hands and feet in the air. I sat reading a newspaper, floating on the water. It was great fun, but murder on cuts—the heavy salt concentration makes the water sting like hell. When we got out, we were very relieved to find fresh water showers on the beach. Otherwise, as we dried we'd have developed a crust like a salt-baked fish in a restaurant.

One of our guides through ancient Jerusalem was a young woman named Ruth Dror. She spoke several languages fluently, was very friendly, and besides all that she was quite a dish. As it happens I was between wives at the time—my wife Jasmin had died some eighteen months earlier—and so I was both flattered by her overtures of friendship and in a position to enjoy them. One day during the conference, she said to me, "If you don't have to rush back to the States, maybe you'd like to come to the Weizmann Institute with me." Would I? Quite apart from Ruth's charms, the Weizmann Institute, in Rehovot, was considered a mecca for scientists throughout the world. This research institute/think tank reflects the Israelis' great respect for intellectual endeavor and liberal support for fundamental science. It turns out that Ruth's father was the chief financial officer for the Weizmann, and he was having a special gathering there.

I accepted immediately.

The Institute itself was set amidst orange groves, and my first morning there, I awakened to the gorgeous perfume of orange blossoms wafting in through the windows.

The reception itself was a new experience for me. As an academic I was used to being surrounded by people whose wealth was in their intellect. The Weizmann affair, by contrast, was set up as a 'thank you' to the Institute's most generous donors. Needless to say, all of the invitees—except me—were extremely wealthy. Since wealth often breeds snobbery I was concerned about the welcome I'd receive, but everyone there seemed solely interested in talking about the progress of science and its benefits to mankind. I was very much in my element and found everyone there extremely pleasant. I almost—but not entirely—forgot that my fellow attendees were simply dripping in doubloons. When I learned that an airport workers' strike had shut down the airport and effectively stranded me in Israel, one of the invitees—a member of the Bronfman family, owners of major distilleries in the U.S. and Canada—offered me a lift out of the country on his private plane. As it turns out, Ruth Dror was eventually able to get me on a 747 flown in by KLM to pick up stranded tourists.

I later spent a three-week sabbatical at the Weizmann, but without the wealthy donors in attendance, I no longer felt like Cinderella at the ball!

Chapter Nineteen

Extreme Skiing at Kitzbuhl . . . But Not on Purpose

Expert skiers will tell you that people who get injured are skiing beyond their skill level. Because I often traveled alone during the 1970's, when I got a chance to ski I didn't have anyone to warn me whether I was skiing beyond my intermediate skills or not. Because of this, at least once I came a grasped stripling away from death on the ski slopes.

During my round-the-world lecture tour in early 1975, I found myself with a free weekend at Easter time in Germany. Having never skied Europe, I decided to head to the famous ski resort at Kitzbuhl, in Austria.

I was relatively confident, having skied all over North America, from Snowbird and Alta in Utah, to Vail in Colorado, Banff in Canada, and Stowe and Killington in Vermont. When I got to Kitzbuhl there was nothing to suggest that skiing there would be substantially different from anything I'd experienced.

The town itself was a Tyrolean dream, and the day I arrived it was sunny and beautiful. *Too* nice, perhaps, because all I saw for vast stretches were green fields; snow was only visible at an extreme distance. At a local ski rental store I asked if skiing were possible in this weather, and the attendant vigorously nodded, "Oh, yes, there is snow around. You have to take the ski lift to the top of the mountain to find it."

I was game. When I got off at the top of the gondola ride up the mountain, it was soon obvious that European skiing was *very* different from anything I'd experienced. In the States and in Canada, ski slopes are broad and well-marked. You know immediately if you're skiing on a slope that's easy—they're marked with green circles—all the way through to triple black diamond for the most difficult.

By contrast, what I faced at Kitzbuhl were hairsbreadth trails with the mountain going straight up on one side, and a sheer drop on the other. There were no signs indicating the difficulty of the trails to come. I was quickly to find that they were beyond triple diamond. They were life-threatening.

I took off down one trail, occasionally seeing signs in German which were unfathomable. In retrospect they should have said something like "Stop now! Turn back! Save yourself!" The trail occasionally opened into patches of woods and fields, replete with spring crocuses and daffodils. These fields taught me that snow skis do, in fact, work on wet grass, especially if you get a flying start.

One field fed me into a snow-covered trail, which barely held my two skis side-by-side. There was a sheer wall of ice to my left and on my right, a precipitous 45-degree drop down the mountain.

Suddenly, the trail took a hairpin turn to the left. Unfortunately, I went straight. Had I been in a James Bond movie, my parachute would have opened at this moment, unfurling the Union Jack, and I would have drifted to the ground. Instead, I instinctively flung out my arms and legs, and in seconds hit the ground spread-eagle. All ajumble with my ski poles and dangling skis, I shot down the mountain as though fired from a cannon, going faster and faster.

In the curious calm that accompanies disaster, as I tumbled I guessed that this steep incline might continue to the bottom of the mountain, and if I stayed balled up I might well roll all the way down, bruised and embarrassed but alive.

All at once the terrain changed. I was flying through woods, trees incongruously growing on the steep slopes. More by luck than plan, I reached out and grabbed a stripling. It broke my fall and I snapped to a stop.

As I looked around, blinking and gasping, I realized that this was a stroke of good fortune only in the most general sense. I was clinging perilously to a fir tree so small it made Charlie Brown's Christmas tree look like a giant redwood, with no one apparently in shouting distance and no clue of how to save myself. It was only after a few minutes that I noticed a ledge nearby. I figured out that if I could release the straps from my skis and poles and wedge them uphill of the tree, I could kick my toeholds into the icy snow and climb to the ledge. It took the longest ten minutes of my life to clamber up to the plateau, with my heartbeat dancing a rumba in my head.

When I pulled myself onto the ledge, I peered back and realized that if I'd continued sliding—my original plan—I would have gone over the edge of a one hundred-foot cliff.

My brief reverie was interrupted by another skier, who lurched off the hairpin trail exactly as I had done, and likewise grabbed a fir tree, this one about six feet below my ledge. I went immediately from being in desperate need of rescue to acting as rescuer. He was Italian, and since he didn't speak English and I spoke no Italian, our only means of communication was schoolboy German. But that didn't much matter, because saving ourselves didn't require a lot of conversation. When he reached the ledge, we worked together to painstakingly climb to a rescue hut. When we safely made it back down the mountain, it was clear that neither of us, in any language, intended ever to ski again.

That evening, I sat in the bar at my hotel, recounting my near-death experience to a group of skiers new to town who were inquiring about ski conditions. Just as I was renewing my vow never to ski again, a young woman sitting a short distance away at the bar broke in, laughing, and said, "I am a ski instructor. I've got a children's ski class in the morning. If you'd like to join us, I can show you the better ski slopes."

I took her up on her offer. Fortunately, it snowed overnight, and when I awoke Kitzbuhl looked like a classic winter picture-postcard. True to her promise, I found the young woman and her ski class, and enjoyed one of my best days of skiing ever, free of any risks to life and limb.

Chapter Twenty

Big Bangs

As a boy, the one thing that had sparked my interest in chemistry was the prospect of making explosives. Fireworks, stink bombs, incendiary devices—the possibility that I could actually make these wondrous things myself was irresistible.

When I got into academia I found that I hadn't been alone in this. A fascination with explosives propelled many of my colleagues into a career involving chemistry. There were often dangerous experiments taking place in the labs at Case Western when I worked there. From my office I routinely heard all kinds of bangs and thumps from minor explosions. But in the fifteen years I worked at Case, there were only three explosions of note.

The first occurred in the late 1960's, in the polymer chemistry lab in the basement of the chemistry building. It was serious enough to cause a student to lose a foot—or, more precisely, *his* foot.

There were two students in the lab at the time. They were working on a senior thesis project. After the explosion, the two students claimed that an old storage bottle of dried-out polymer suspension had fallen off a shelf and exploded. This was at least theoretically possible, but if it was true then both the University and the students' supervisor, Professor Irv Krieger, would be liable for negligence.

Krieger suspected something different had happened. He collected samples of the white residue left from the explosion. If the students' explanation reflected what really happened, there would be evidence of peroxides. Instead, Krieger found that the samples were heavily laden with mercury.

When he confronted the students with this, they immediately admitted that they were, in fact, making mercury fulminate, a high explosive generally found in detonators—which they intended to carry, on a *commercial airline flight,* to Colorado!

The process they were using was actually quite simple, but *extremely* dangerous. They dissolved mercury metal in *aqua regia* (a mixture of pure nitric and sulfuric acids), and separated and dried the white residue. (It's this residue that Krieger scraped up from the floor.) They dried the residue in a drying oven and transferred the contents to a jar. It was a painstaking process, which they carried out in increments of 10 grams. They wanted at least a hundred grams for their plane trip to Colorado.

It turns out that on one of the trips from the drying oven to the jar, one of the students dropped the crucible on his foot, and BOOM! Good-bye foot.

If it is true that nothing is good or bad except by comparison, this student's lost foot was a blessing. If the two had actually made it onto that plane to Colorado, carrying a jar of mercury fulminate, the result could have been catastrophic.

Krieger's lab was ironically also involved in another serious explosion, several years later. The bangs, rattles and noxious odours of minor explosions often reached my office, which was one floor up from the lab. But on this particular occasion, I was in the corridor outside my door talking to a colleague when a particularly loud explosion emanated from Krieger's lab. We probably would have ignored it, except that the explosion was closely followed by yelling and screaming. The two of us ran downstairs. We found dense chemical smoke billowing from the lab. Just as we were preparing to plunge in and look for survivors, we saw Krieger emerging, dragging a Chinese graduate student. It looked like a vision of hell. The student, though conscious, was holding his right arm in the air, with a mass of pulp where his hand had been.

I immediately raced into the lab to look for other victims. Not far inside the lab, I tripped over an unconscious student and dragged him outside. Fortunately for him, he ruptured an eardrum but was otherwise unhurt.

It turns out that sheer stupidity caused the explosion. The Chinese student had been working with sodium metal, which is *very* reactive. Recognizing that he had to keep the remaining metal in an inert solvent, he chose carbon tetrachloride, which used to be used as a cleaning solvent. Unfortunately, sodium reacts violently with anything containing chlorine atoms. This is something that every freshman chemistry student knows, but in this case it was overlooked, and—bang!

In contrast to these two tragic explosions, the third is one I remember with some amusement. One of my colleagues was interested in the structure of cellulose. Cellulose is a biological polymer. It is very common in nature, being the basic fiber found in wood, paper and cotton. My colleague wanted to identify the exact arrangement of glucose (sugar) subunits and the atoms in them by way of x-ray crystallography. And that meant he needed crystals. My research group had already shown that if you used the right mixture of solvents, you could produce crystals of most biopolymers, but cellulose was very stubborn and didn't dissolve in any normal solvent mixture. This spurred my colleague to use more and more exotic solvents, and in this case "exotic" meant "dangerous."

He finally set up an experiment where cotton and the solvent hydrazine would be mixed in a platinum reaction vessel. This was a highly volatile mixture and my colleague and his students used extreme precautions. The reaction was carried out in a fume hood, in which the fumes of a reaction are vented to the outside, and a mesh-covered, safety-glass window protects the experimenter. The fume hood was on a marble benchtop, with several barriers of protection. The graduate student conducting the experiment took the further precaution of evacuating the lab before he started to work.

Nonetheless—KABOOM! It was one of the most powerful explosions our poor Olin Building had ever suffered. The fume hood disintegrated. The marble bench top shattered. The rest of the lab bore the scars, flotsam and jetsam of a major accident.

The explosion was serious enough to attract the media. Within minutes, local TV reporters were on the scene to interview the graduate student, who had fortunately emerged unhurt. I stood by, listening to the exchange:

Reporter: "What exactly caused the explosion?"

Grad student: "Well, I was dissolving cellulose . . ."

"What's that?"

"Sort of like gun cotton. I was dissolving it with hydrazine . . ."

"What's that?"

"Sort of like rocket fuel."

"What precautions did you take?"

"The normal ones. I was lying on the floor, watching the reaction through a mirror."

In fact, the accident was traced to the negligence of the outside contractor who'd built the reaction stirrers, which are metal beakers made of platinum that are specially welded. Or at least, they are *supposed* to be specially welded. Despite explicit instructions to the contrary, the contractor had used a soldering mixture, and this in turn caused the explosion.

So in spite of the inherent danger of the experiment, the accident was caused by simple stupidity—as they often are.

Many years after the explosion, when I entered venture capital, I mentioned this story while touring a young English company called Oxford Glycosystems. A few years after my visit, I ran into the president of the company. He said, "I never told you this, but that story you told my scientists about the explosion in your lab saved their lives." Delighted, I asked how, and he responded, "After you left, they looked at each other and realized that even though they were about to do a similar experiment, they hadn't paid any attention to the reaction stirrers. The ones they were going to use would have exploded, just like the ones you told them about."

I don't know why I told that particular story to that particular audience. If you believe in Fate you would say that it wasn't their time to go. All I know is that a story I've always regarded as a simple anecdote now has a whole new meaning for me—and a few very fortunate scientists in England.

Chapter Twenty-One

Fish out of Water at Harvard

The fall of 1970 was the low point of my life. My wife Jasmin's mental state was rapidly deteriorating. In August, we sent Kimm and Keir to live with their grandparents in England. In November, Jasmin died.

For the first time in the ten years since we had arrived in the United States, I was alone. Work was, as it always had been, a salvation for me. But it was obvious to people who knew me that I was not a happy man.

A colleague of mine, Elkan Blout, recognized this. I knew him through our research in similar fields, and also because he had recruited one of my graduate students as a post-doctoral fellow. At the time, Elkan was head of the Department of Biological Chemistry at the Harvard Medical School. He suggested that I take a leave of absence from Case Western, and go to Harvard as a visiting professor for a year or two. I immediately leapt at the chance. It seemed like an excellent opportunity to try to get everything back on track. (It did in fact turn my life around. I emerged after a year with my first major biological chemistry book well under way, and went back to Case revitalized.)

I didn't want to live alone at Harvard, so instead of taking an apartment, I lived on campus with the medical students at Vanderbilt Hall, a short walk from the Biological Chemistry Department. It was a tremendous social and intellectual shock, just the kind of jolt I needed. Although I was very depressed at first, my spirits lifted as I got to know people. I started dating a TWA stewardess who flew frequently to London and Paris. She brought me back all kinds of English goodies—sweets mostly—and I loved that.

In the dormitory itself, I was in every way a fish out of water. It was the height of the Vietnam War, the flower power era, and everybody else in the dorm was single and at least ten years younger than I was. They were also wealthy, something I most definitely was not.

At first they treated me as something of a curiosity. I didn't help matters by refusing to wear blue jeans, the uniform of the time. Until fairly recently I viewed blue jeans as inappropriate wear for anyone except a laborer, but I eventually found that it is not possible to live in the United States and not wear jeans. I think it must be a law.

In my favour I *did* play the guitar, and that gained me entry to more than one student party. As I met people I noticed immediately that every person there was not just an excellent student, but also very accomplished in some other activity. The fellow with the room opposite mine was on the Olympic sword fencing team. His neighbor had been on the first U.S. ping pong team to visit China. Every night, in the common room downstairs, someone would sit down at the piano and rip off an extraordinarily difficult classical composition from memory. There were tennis courts in the quadrangle, and I had brought my racquet hoping to get in a game or two. I strolled over there a few times and on each occasion watched the best amateur tennis I have ever seen. I never took my racquet out of its cover.

Another revelation to me were the courtship rituals of the students. I had grown up as an only child and attended solely or predominantly male institutions from the time I was eleven. My grammar school headmaster at Kings Norton had huffily stated, "Plenty of time for the sexual sideshow later."

It was to be some time before I got to an eyeful of that sexual sideshow for myself. In fact, one of my first dates at Nottingham was memorable, but not for romance.

My date was a pretty blonde student named Mary Lloyd. We went to a movie in downtown Nottingham. We were slightly late, and the old movie theater was packed. The movie had started, and the only two adjacent seats left were at the very end of one of the middle rows. Getting to our seats meant that several people had to stand up and let us pass.

We got to our seats, sat down, and leaned back, just as everybody else in the row was settling in. There was a mighty CRA-A-A-ACK as the supports at the ends of the row fractured and collapsed. Our whole row of people flipped backwards, winding up in the laps of the people behind us.

Because of the ensuing pandemonium, the movie was stopped and the house lights came on as people extricated themselves from the fray. Mary and I spent the rest of our evening at a local pub, on what turned out to be our last date.

Shortly after that I met my first wife Jasmin, and that ended my dating days. So I hadn't seen much of the sexual sideshow that my headmaster had mentioned.

I made up for it with what I witnessed at Harvard.

Next door to the Medical School was Simmons College, an all-female school. This turned out to be superficially synergistic, since the girls were looking for medical student husbands and the med students were looking for sex.

While strumming my guitar at student parties, I was a silent witness to many pickup routines. I was amazed at the frankness of these opening gambits. I remember hearing one student tell a woman, "I saw you at a party the other day and thought that you were pretty ugly, but you might be a pretty good lay."

The most brash picker-upper I remember was my neighbor in the dormitory. He was a first year med student who had already written a best-selling book called *How To Survive in New York,* which he was working at converting into an off-Broadway play. His other ambition seemed to be sleeping with a different girl every night. At parties, he basically had the attitude that it was a privilege for any Simmons student to sleep with him.

Our dorm walls were very thin, and I remember an exchange I heard between him and one of his nightly enamoratas. She protested, "But you won't respect me in the morning." He responded, "Respect you? I hardly even know you."

He wound up number one in his class at the end of the year. So he scored in more ways than one.

Chapter Twenty-Two

Giant Rats

It is very easy to get an inflated opinion of yourself when you are a tenured professor at a significant private university. You expect, and receive, a lot of respect. That's how it was for me, at Case Western, from the time I became a tenured full professor in 1971—until my daughter Kimm arrived on campus, in the late 1970's.

As is true of most universities, one terrific fringe benefit at Case Western was free tuition for relatives. When Kimm was accepted at Case, I thought it was wonderful. But her practical joking nature occasionally made me reconsider my attitude.

Once, I was walking through campus with the noted Nobel laureate Paul Flory. Kimm ran up to us, breathless, and did not let on that she was my daughter. "Dr. Walton!" she exclaimed. "I've always admired your work! May I have your autograph?" Flustered, I did, in fact, sign the paper she thrust in front of me.

Another of her practical jokes involved a class I taught. It was a graduate biomolecular chemistry course called "Biopolymers." It's worth pointing out that the students in this class felt very lucky to be there; to get into the graduate program at all was exceptionally competitive. There were four or five applicants for every opening. Students who took my Biopolymers class were very anxious to do well and make a good impression.

Kimm showed up at the beginning of class one semester, and sat down among the twenty or so genuine grad students. Because in theory she was entitled to audit any class, I decided not to make a big deal of this appearance. As I started to lecture, the real students began furiously taking notes. Kimm sat, nonchalantly, looking entirely bored. This garnered a few stares from her neighboring students.

After a few minutes, she got up, strode to the front of the room, and asked me: "Does anybody else teach this class?" When I said "No," she walked out, leaving a stunned silence behind her.

On another occasion, I was reading through mountains of applications to the graduate program. Sure enough, there was an application from a "Kim Walton." Very clever, I thought, and without paying much attention to the application, I wrote a very snotty rejection.

A few days later, I complimented Kimm on her application. She looked really surprised, and asked, "What application?" It took her several minutes to convince me that she had not, in fact, applied to my program, and that there genuinely *was* an applicant named "Kim Walton"—one whom I had just astounded with my rejection, since *he* was a straight-A student. One of her finest jokes on me was not even her own creation.

After a couple of years of being the brunt of these kinds of humorous assaults, I decided it was time to get my own back. I planned an elaborate hoax. Kimm had a pet white mouse, which she kept in an aquarium in her room. One day, over breakfast, I casually asked her: "Is it all right if I take your mouse to the lab to do an experiment on it?"

Aghast, she responded, "No way!" just as I had expected.

"Oh, come on," I pleaded. "The experiment won't hurt it."

After several days of similar entreaties, she finally gave in. I told her I had been working on a genetically-engineered growth hormone at the lab, and I wanted to inject her mouse with it to see if it worked. "What will happen to him?" she asked me suspiciously.

"At most, he'll grow about fifty percent," I responded.

I brought home a small syringe filled with saline solution, and, in her presence, I gave her pet mouse a shot of "growth hormone."

What she didn't know was that I had spirited home a large white lab rat, and I had it hiding in a cage in the master bedroom. That afternoon, while Kimm was otherwise occupied, I switched her mouse for the large rat. At that point, my wife E.J. and I went out to the movies.

We came home to an uproar. Every light in the house was on. As we pulled up to the garage Kimm ran out of the house, screaming that I had worked a miracle.

"My mouse is three times as big as it was this morning! It's like—it's like—Jiffy Pop! It's huge! Oh my God! You're a genius!" she shouted.

When E.J. and I got in the house, the "giant mouse" had everyone agog. Keir and Kimm were plotting my route to fame. Kristin and Sherri, at that time only little girls, were in wide-eyed disbelief. Kimm's date for the evening, an engineering student named Kirby, was skeptical that the mouse's skull could have grown so quickly, but he was soon shouted down by Kimm's utter belief in my "genius."

I don't know how I kept from collapsing with laughter. I thought I was going to implode. I had to shut myself away for half an hour to laugh it out of my system before I could keep a straight face.

Kimm and Keir stayed up half the night watching this miracle animal. Kristin and Sherri took up residence downstairs, refusing to sleep on the same floor as the "giant mouse." They were apparently convinced that, like Alice in Wonderland, the mouse wasn't going to stop growing. Their fears were not calmed when Kimm told them that during the night, the "giant mouse" had learned to remove the lid from the aquarium—the equivalent of a four-story "mouse house"—and climbed out of the cage, winding up on Kimm's pillow and nibbling her ear.

By breakfast the following day, Kimm and Keir had come to the conclusion that this was such an enormous scientific advance that they were going to call the local newspapers and have them send photographers. They envisioned farm animals from cows to chickens being similarly injected, solving any conceivable food crisis.

Obviously I couldn't let this happen without letting the joke come unglued. While Kimm and Keir were gung-ho science—my influence I believe—the general public was already apprehensive about genetic engineering. And beyond that, any knowledgeable scientist would immediately expose the fraud. Searching frantically for an excuse, I told Kimm, "You can't call the newspapers. Experiments like this are strictly illegal. That's why I needed to use your mouse instead of one from the lab." Disappointed, she and Keir gave up their planned media assault.

I had intended to let the joke go on for a few weeks, and then replace the rat with Kimm's pet mouse, saying that the "effect" of the growth hormone must have worn off. Unfortunately, Kimm's mouse died in the meantime. And she certainly seemed fascinated by her new economy-sized accidental pet. Because it was too large to fit in the mouse-sized wheels and toys in its aquarium, she fashioned large ones out of different pieces of tubing. Since it had a giant appetite—it would get through a box of mouse food in a couple of days—she experimented with different foods, finally feeding the voracious rodent Purina Dog Chow. And she regaled every new acquaintance with tales of her father, the "scientific genius," and his miracle growth hormone. "And he doesn't even get any *credit* for it!" she'd exclaim.

After a year or so, the "giant mouse" died. At that point I figured I would never be found out. But the legend lived on. Some time later,

Kimm was telling one of my graduate students about her giant mouse and her genius Dad. While most of my students would have kept my secret, she happened to be talking to Mark Soderquist, a devout Mormon. He gave me a stern look and said, “Alan—you mean you haven’t told her?”

The jig was up. Now everyone knows that my contribution to the advancement of science won’t involve rapidly expanding mice.

Chapter Twenty-Three

"If We Don't Play God, Who Will?"

Scientists are the true heroes. I like to think I'd say that even if I weren't a scientist myself, but of course I can't know that for sure.

During times of war, scientists can invent synthetic fuels, new weapons, and new materials that decide who wins and who loses. Dramatic stories about the creation of weapons like the atomic bomb are the stuff of legend. But what those scientists did is what scientists have done through the ages—they looked beyond what we know into the vast expanses of what was yet to be, and through the sheer power of thought made the previously unseen, real.

I once had my academic "family tree" researched back to the 15th century. It showed my Ph.D. advisor, and his Ph.D. advisor, and *his* Ph.D. advisor, and so on, all the way back to a scientist in Padua in the 1400's. Think of the scientific world he lived in! My academic "father" to the 30th power lived in a world that believed the Earth was flat, before Galileo and Newton.

We pride ourselves on the technological advances we've made today. What thrills *me* is the thought that through science, there is no question that cannot potentially be answered. There is *no* bigger adrenaline rush than that. It's always seemed to me that this ought to thrill *everybody*. However, the media is always full of stories showing that the general public's awareness of how things "tick" is abysmal.

The public view of science always comes home to me when I'm interviewed by the media. As a professor, biotech company president and venture capitalist, I've been involved in a number of question-and-answer sessions. They always bring home to me the vast gulf between the state of science—and scientific possibility—and the perception of the public.

Once, during a radio call-in show in Connecticut, a young woman asked, "You're into genetic engineering, right?" I said that I was, and she went on, "So why don't you genetically engineer astronauts so they don't need space suits?"

During another question-and-answer session I was once asked, "Why can't we engineer soldiers in Vietnam to be bullet-proof?"

I am never quite sure whether my leg is being pulled by questions like this, but to be on the safe side I usually try to rephrase the question into one that is actually being seriously addressed by the scientific community. So I'd talk about how we were working on genetically-modified plants to provide oxygen and food to astronauts. Or pills soldiers could take ahead of time to neutralize poisonous gas attacks. Or new vaccines to handle rare tropical diseases.

Of course, some people who ask me questions aren't interested in the "gee whiz" aspects of science. Instead, they like to debate the ethics of prenatal gene therapy and other controversial topics. They say that difficult cases are the quicksands of the law; similarly, difficult ethical issues are the quicksands of science. I think most people probably would agree that if we could correct very serious defects like Fragile X syndrome (a devastating mental disorder) while a child is still in the womb, that would be acceptable.

However, the general public would probably squirm at the idea of changing eye colour or increasing intelligence through gene manipulation. My hunch is that people's opinions about issues like these will change over time. Many proscribed procedures today may be acceptable then.

After all, it took the Vatican hundreds of years to accept that the Earth is not the center of the Universe.

XXXXXX

My first exposure to scientific celebrity came at Nottingham. I walked by a lecture hall where Albert Einstein had been speaking. His chalk notes on the blackboard had been preserved. Since I didn't hear the lecture—what a missed opportunity!—I can't recount a witticism from it. In fact one of my favourite stories about Einstein has nothing to do with science.

At a dinner party celebrating his wedding anniversary, Einstein was asked the secret of his happy marriage. "It was easy," the great man responded. "Before my wife and I got married, we agreed that in our marriage, I would handle the big decisions—the shape of the universe, the nature of space and time, those kinds of things. And she," he continued, "would handle the small things."

He paused for a moment, adding: "And in all this time, there hasn't been one big decision."

One of the most brilliant scientific speakers I have ever known is Ephraim Katzir, former President of Israel and a good friend. (Ironically enough, Einstein had been offered the Presidency of Israel in 1952, three years before he died.)

I first met Ephraim in the late 1960's when he was Ephraim Katchalsky, a professor at the Weizmann Institute in Rehovot, Israel. He was a terrific speaker, full of charm and wit.

Of the many hundreds of international talks and lectures I have given, I have always looked forward to speaking after a mediocre lecturer, so I'd seem great by comparison. On one occasion, I was asked to give an after-dinner talk at a scientific conference. To my horror, I learned that I was to speak immediately after Ephraim.

As usual, he gave one of his dazzling performances. With every laugh he got, I sank lower in my seat, dreading the thought of following this star. When I got up and spoke, I did pretty well—or so I thought. But as soon as I was done, the audience enthusiastically insisted that Ephraim get up and speak *again!*

I never thought to ask him how he became President of Israel—the combination of science and politics isn't very common, after all—and I certainly had the opportunity, because I visited him several times in Israel after his election. He had maintained his strong interest in science, and often invited scientific colleagues, like me, to cocktail parties at his official residence. At one of our meetings, Ephraim's involvement in science and politics came into close proximity.

It was a small seminar in Ayelet Hashachar, a kibbutz near Mount Herman and close to the Lebanese border. The time was late 1975, when Israel was at war. From our vantage point, we could watch the Syrian/Israeli air combat and hear the distant thunder of guns. Our group—numbering fifty scientists or so—had about two dozen Israeli soldier guards, each equipped with an Uzi machine gun. It was *very* exciting!

Around that time, Henry Kissinger, Secretary of State under Presidents Nixon and Ford, had been carrying out his famous "shuttle diplomacy" between Egypt and Israel. His efforts eventually culminated in a famous peace treaty, the Camp David Accord, signed during the Carter presidency.

With Ephraim's role on the world stage, I knew he would be familiar with Kissinger, and I asked what he thought of him. He related the following story. "Towards the end of his travels back and forth between Cairo and Tel Aviv, I asked Kissinger how he could handle all of the back stabbing that he encountered in both capitals. 'It's easy,' he deadpanned. 'I am a professor at Harvard University.'"

xxxxxx

My career in biotechnology wouldn't have been possible without Francis Crick and James Watson, because without them, there wouldn't *be* biotechnology. They discovered the structure of the DNA molecule, which opened the floodgates for modern biotechnology and the molecular genetics revolution.

Their discovery was brilliantly chronicled in Watson's book *The Double Helix*. It tells of his experience as an American post-doctoral fellow at the Cavendish laboratories at Cambridge University, where Francis Crick was the lab director.

I first met James Watson when he was in his seventies, and was acting as the Director of the Cold Spring Harbor Labs on Long Island. I had the honor of awarding him the National Biotechnology Medal for lifetime contributions to the biological sciences. After that, we met several times, chatting always about science and philosophy.

As the co-discoverer of the structure of DNA, Jim is the product of many years of scientific adoration, which has gone to his head. He is a very curmudgeonly, opinionated person who calls them as he sees them. Among other things, he is an atheist, a big supporter of assisted suicide pioneer Jack Kevorkian, and a believer in voluntary euthanasia for those who are drastically disabled. His frank comments have often upset people who have relatives with genetic defects. But he is inescapably a provocative thinker and speaker, still producing novel ideas and staying on the cutting edge of science.

In 1997, a couple of years after I met Jim, I met his colleague and mentor Francis Crick. By then, Crick was 80 years old, and still active at the Salk Institute in San Diego. He had a marvelous story about to tell Watson.

"You remember, of course, when they cloned that sheep, Dolly?" he asked. How could I miss it? They cloned her from the epithelial cell DNA inserted into a vacated sheep egg cell. "A week after the experiment was announced, Jim was interviewed by the press, eager to grill him about the touchy subject of human cloning." Crick went on to relate that one young female reporter had been giving the great man a hard time about science in general, and she asked him, "Aren't you scientists playing God?"

Watson's crusty reply: "If we don't, who will?"

Chapter Twenty-Four

Presidential Advisor

As an immigrant I'd always been fascinated with the glamourous goings-on in Washington, D.C. Power lunches, glittering embassy parties, fingers on the pulse of history . . . it all had a special patina to me.

I knew that it was possible for scientists to get involved in politics, because I'd met a former science advisor to President Eisenhower when I was a newly-minted professor at Case Institute, in 1963. This fellow had just arrived from Asia and was visiting Case on his way back to Harvard. I might have forgotten him entirely except that he told me a wonderful story.

It seems that two of his friends who'd been working in South Africa were on their way back to the States via Hong Kong and Tokyo. At the time, the thing to do with a short stopover in Hong Kong was to be measured for a custom-made suit. It seems that you could get one of these suits for about one-tenth what they'd cost you in the States.

So one of these chaps ordered a suit his first day in Hong Kong, intending to pick it up the following day on his way to the airport. Unfortunately, he and his friend were running a little late. They arrived at the tailors' with only time enough to quickly don the suit and leave. It looked great. The lucky purchaser threw his regular clothes into his suitcase and jumped into a cab for the airport.

During the cab ride, he soon began to notice problems with the suit. For one thing, the pockets were artificial. More curiously, the cotton thread holding his trousers together was not up to the job. Gradually both the inside and outside seams started to unravel from the cuffs upwards.

By the time he reached the airport, the seams up to his knees on both legs were completely undone. Late for his plane, he ran toward the gangway with the seams quickly disappearing. When he mounted the stairs to enter the plane, his trousers consisted of two pieces of material, front and back, held together by his belt, like some kind of modern-day toga.

He tried to keep his modesty by holding his briefcase in front of him. Nonetheless, he was the last passenger to board, and a hundred pairs of eyes were immediately riveted to him.

The stewardess, with incredible presence of mind, quickly wrapped him in an airline blanket and bundled him to his seat. After takeoff, she asked the embarrassed fellow to remove the remainder of his trousers. She and one of the other stewardesses made rapid repairs, doing a better job of sewing his pants than the tailors had done.

In these days of three-peanut airline meals, those times seem very long ago indeed.

I got involved with politics in a roundabout way. In 1976, I was chairman of one of the divisions of the Biophysical Society. I volunteered to talk to the political candidates about their views on promoting science. One of the candidates I contacted was Jimmy Carter. He didn't respond to my letter, but as luck would have it, I met him shortly thereafter.

My daughter Kimm was volunteering for Carter's primary campaign as a high school project. She convinced me to spend fifty bucks and attend a fundraiser for the governor, to be given at a downtown Cleveland hotel. After his requisite ten-minute pep rally speech, everybody lined up to shake hands with him. While most of the people he greeted gave him a hearty "good luck," I thought it was a perfect opportunity to ask him why he hadn't answered my letter.

That's exactly what I did. His toothy grin wavered for a moment, and then he responded, "Where and when did you write to me? What did you write about?" When I said I'd sent the letter to his Plains, Georgi,a headquarters, he answered, "I haven't been back to Plains in a couple of months. But I'm very interested in science policy. I haven't put together a policy plan yet." He went on to ask me to contact a friend of his in Washington, who was starting to think about scientific issues on Carter's behalf.

His friend turned out to be British, and we got along quite well. He put me onto a Democratic Party staffer in Washington, a real "doer," high energy, bright, with an immediate grasp of the fact that Carter had no science policy, wasn't well-versed on the issues, and was potentially vulnerable as a result. This staffer asked me to submit a list of fifty or so prominent scientists from various fields as potential scientific advisors to Carter. From this list seven science advisors would be chosen, and they would put forward issues and potential solutions relating to science policy.

It turns out that I wasn't the only one asked to supply such a list. In fact, heated competition sprang up between the Democratic Central Committee (DCC) in Washington and Carter staffers in Georgia. Lewis Branscomb, head of R&D for IBM and a friend of Carter's, was the leader of the Georgia group and he was coming up with his own list of potential advisors. The whole thing was incredibly disorganized.

In the meantime, then-President Ford's advisors were starting to lob a few scientific hand grenades at the Democrats. There were several occasions when I was called on, in the middle of the night, to respond to these issues.

In the end, Branscomb submitted his list to the National Democratic Committee in Washington, and they passed it on to me for comment. To my astonishment every single person Branscomb had named was a physicist—hardly a balanced representation of the scientific disciplines! As a result Branscomb's and the DCC's groups were merged under Branscomb's leadership. I wound up being one of the seven advisors.

We spent a lot of time developing a list of issues and our stances on them so that Carter had a jumping-off point for forming his own positions. There were about sixty issues on the list, ranging from reaction to national emergencies like hurricanes and floods, all the way to expenditures in high-energy physics.

As it turns out, of our sixty issues Carter disagreed with us on only two. One involved NASA. We had encouraged increased spending on the space program, believing that spin-off technologies more than justified all expenditures on NASA. Carter told us that although he was personally in favour of spending more on NASA, the American people regarded sending rockets into space as a giant (and very expensive) fireworks display.

The other issue on which we disagreed involved energy. At that time, in 1976, the U.S. was still smarting from the oil embargo of the early 70's, with very high gas prices and oil shortages. We encouraged exploration of alternative energy resources. Carter instead argued that the American public was frightened of nuclear power and radioactivity, so no alternative policy was developed.

The way the committee worked was to meet in subgroups. Carter met with the subgroups briefly, and then finally with the whole group for an hour or so. I found him very personable. Unlike most other politicians I've known, he listened and made decisions based on the evidence rather than dogma. In person he was quite charismatic. I often thought it a shame that this quality was lost on television—something I understand is true for quite a few politicians.

One of the benefits of working on the committee was an invitation to Inaugural Balls. There were seven, and I got invited to two. While some of the Balls were for high-powered government insiders, the ones I attended were focused more on the Party faithful. Nonetheless, I got to meet quite a few very interesting people. One of them was the Governor of North Dakota. When the party was winding down, he discovered that his limo driver had gotten drunk. So I ended the evening helping the governor bundle his chauffeur into the back of his limousine—and the governor piloted his own limo back to his hotel!

Some of the issues that we worked on in 1976 and 1977 later led to important governmental policy changes.

One was the Dole-Bayh Bill of 1981. It allowed universities to exclusively license technology. That was a major boost to biotechnology, because without an exclusive license, no startup tech company can hope to attract capital—the competition of other companies doing the exact same thing would squelch any investor interest. Government laboratories followed suit in 1986, which directly led to the creation of such groundbreaking companies as Human Genome Sciences, which is using the human genome to produce a new generation of pharmaceuticals.

One of the most fascinating aspects of my involvement with Carter is something I never actually saw. I heard a rumor that a list existed, naming a few hundred scientists who would be protected in the event of a nuclear attack by assembling underground in nuclear shelters—very James Bond-ish. Although I never got my hands on this list, I was told that the criteria for inclusion involved choosing the scientists in various fields with the most number of scientific publications, as listed in the Scientific Citation Index.

Naturally, I went straight to the Scientific Citation Index, and scanned it to see how likely I was to be included on the "saved" list.

I convinced myself I was!

Chapter Twenty-Five

"I Learned About Flying from That"

Flying Magazine has for years had a column called "I Learned About Flying from That." In it, private pilots recount their own stories about sticky flying situations and how they got out of them. There can hardly be a pilot who doesn't have a story to contribute. As in every other facet of life, in the air—things happen.

I got back into flying in 1973 after I returned from Harvard to Case Western. I found out about a small club that owned a Piper Colt, an old two-seater plane. It looked to be much easier to fly than the Chipmunks I'd flown in the RAF Reserve.

The way the Piper club worked was that for a small fee, you could own a share in the plane. On top of that, for six bucks an hour you could get instruction at a nearby grass strip. What a deal! I signed up immediately.

It turns out that with my experience in the RAF Reserve, I was ready to fly solo after three hours of instruction. Unfortunately, shortly thereafter fall and winter set in, making the grass field unusable. It wasn't until the spring of 1974 that I got to take the plane out for another spin. And, as luck would have it, I was the first one in the club to use the plane after its three-month layoff.

With all of twelve hours of private flying experience, I checked out the creaky old plane and took it up for a few trips around the circuit. The circuit at any airport is the rectangular pattern you use to practice landing. The runway forms one of the long sides of the rectangle; the other long side, on the right or left side of the runway depending on which way the wind is blowing, is called the downwind side, and the short sides are crosswind.

As I flew the circuit, everything seemed fine—until I got ready to land. This was slightly tricky in the Colt. Normally, planes have flaps which allow for a controlled angle of approach when you land. The Colt didn't have flaps. That meant that in steep approaches you had to use a maneuver called "side slipping," in which you lower a wing into the wind and put on opposite rudder, to keep the plane straight. What this does is to decrease the lift surface and make the plane descend more quickly. It doesn't sound easy, and it's not. To add to my difficulties, the grass runway was about as short as a public airfield is allowed to be. I didn't have a large margin of error.

I didn't expect to nail my landing on my first attempt. And I was right. I was way too high on the final approach. I flew a circle around again, and started much further back for my final descent. To my great surprise, I once again sailed several hundred feet over the field. I started to think I'd lost my touch. But it had only been six months since I'd been in this plane. I couldn't be *that* rusty!

On the third trip around, I discovered the problem. The throttle was sticking open, giving the plane too much power to land. I tried to alternately open and close the throttle to cut the power, but that didn't work. Instead of cutting the power to the engine entirely, it just partially reduced it, kind of like when you have a throttle that sticks open on a lawnmower.

Finally, I just cut the engine, doing what's reassuringly known as a "dead stick" landing.

When I got out, there were pilots hanging around on the ground who'd been watching me. They were laughing when I walked up, one of them telling me, "We thought we'd have to shoot you down."

Now, with a rented plane, this never would have happened. The FAA mandates that rented planes go through a thorough once-over every 100 hours they are flown. Privately owned planes—like our Colt—only require an annual inspection. With more frequent checkups, the throttle problem might have been detected—and my kamikaze imitation avoided.

I've flown rented planes ever since.

XXXXXX

Although planes have many gauges and switches, flying itself is pretty simple. But people who don't fly don't realize that, which allows a lot of room for pulling legs.

In the 1970's and 1980's, my son Keir frequently came along for the ride when I took out planes for a spin. He particularly enjoyed it when I gave him things to do. Since planes have dual controls—the co-pilot can take the wheel—I often let him fly the plane. While there was nothing he could help with on takeoffs, when I was landing I would often have him set the flaps. This was accomplished by moving a switch between the pilot's and co-pilot's seats.

On one occasion, when we were getting ready to land at an Ohio airport, I gave Keir the order, "I need twenty degrees of flaps."

Keir reached down to set the switch, and in a scene directly out of the *Three Stooges*, the damn thing came off in his hand. "Dad! Look!" he cried. His face was ashen as he showed me the switch.

Now, I *could* have instantly reassured him that it didn't matter. In fact, you can land a plane without altering the flaps at all; it's one of the first things you learn in pilot training. But I couldn't resist putting him on, at least for a moment.

"Oh, no!" I said. "How am I going to land the plane?"

Keir was convinced we were going to crash. But of course we didn't.

When we got home, I overheard him telling E.J., "When that switch came off in my hand, I saw my whole life pass before my eyes." Since he couldn't have been more than about twelve, it couldn't have taken very long. I immediately felt guilty. The poor kid had honestly thought this was the end.

It was a few months before I persuaded him back into the cockpit again.

The cosmic payback for my joke on Keir came in 1989, when I had two flights that made up for years of smooth sailing.

In January, my youngest daughter Sherri had to make a documentary for one of her college classes. She wanted to make an instructional tape called "How to fly a plane." Her plan was to watch me do a preflight check and explain the principles of flight, and then she would explain them on film. Then we would go for a short flight and film that as well. The videographer would be her friend—and later husband—Paul.

I rented a Piper Warrior for the occasion, from the local Fixed Base Operator (called an "FBO"), which is essentially a Hertz rent-a-plane. Unfortunately, we hadn't chosen a brilliant day to fly. It was very cold, with snow showers in the area. It was marginal "VFR" (visual flight rules) weather at best, meaning that the weather was clear enough to fly "visually," without having to be licensed to fly on instruments only (which is called being "instrument rated").

The snow showers meant that conditions could quickly deteriorate to the point where VFR flying was impossible. Although I was instrument rated, I hadn't flown on instruments in a long time and didn't feel comfortable doing it now. On top of that, the FBO warned me that the directional gyro on the plane was not working well.

The directional gyro, also known as a D.I. or direction indicator, gives a very accurate indication of which direction you're flying. It is very steady, and when you use it properly, it's accurate to about one degree—very precise indeed. A compass, on the other hand, swings with the motion of the plane—kind of like one of those bobble-head dolls you sometimes see on the back ledge in people's cars. It's not terribly accurate, and to make matters worse it is influenced by any metal or magnetized objects nearby.

Nonetheless, for a single flight around the traffic pattern—basically the equivalent of going around the block—they said that the lack of a directional gyro wasn't a big deal, "You shouldn't have any problem." I was more than a bit dubious about renting a plane that was not fully operational, and I was vaguely aware that this was against certain flight rules. However, Sherri was very enthusiastic about the flight, so we went ahead with it.

The preflight filming went well. There certainly wasn't any background noise to interrupt us. The marginal weather had chased away any other potential flyers that day. While I ran the preflight check, I had the radio tuned to weather reports. The news was going from bad to worse. The cloud ceiling was lower than acceptable for a VFR flight. I told Sherri, "With the weather like this, we should probably just start the engine and taxi around the airport." I went on to assure her that the visual effect on video would be very much the same. But the look on her face suggested this was a big letdown, and I felt like a bit of a weenie disappointing her. At that moment, the weather report on the radio suggested that the low cloud had lifted slightly—just enough for a VFR flight.

The three of us hopped into the plane and took off quickly. I called the tower to tell them that we would make a single circuit of the field. The Danbury field has hills to the east and south of it, so I was relieved when the tower assigned us to the north/south runway, taking off toward the south.

The flight was entirely uneventful until I turned crosswind for the final approach. A snow shower descended instantaneously. The visibility, which had been a comfortable five miles, suddenly plummeted to zero. The tower called to me, "Weather has deteriorated to instrument only conditions." Well, duh. I immediately dredged up what I remembered of instrument flying, and felt that even though I couldn't see a bloody thing, knowing the location of the field would give us the chance to land.

I had to make one more turn in order to land safely. I had to rely on the swinging compass because just as I'd feared, the directional gyro was useless. As I timed a turn onto what I hoped was the final approach, the tower called and asked, "Where are you?" The truth was that I didn't really know, except that I was about a mile away from them.

I responded, "I'm not sure. I can't see anything through the snow."

They suggested, "Look downwards." It was a good idea, since by looking straight down, I could see a freeway intersection off the end of the main runway.

The tower gave me a course to fly, but relying solely on the wobbly swinging compass, I could only approximate it. I had only a moment to think of the irony—essentially lost within a few hundred yards of the runway!

Sherri, having flown before, had a pretty good idea of what was going on. Paul had never been in a small plane. He was sitting, essentially catatonic, in the back seat.

I decided that the best thing to do would be to conduct the approach as though everything was normal. That meant stabilizing the plane at 75 knots with half flaps. The amount of flaps you use to land varies from airplane to airplane. More flaps mean a steeper angle of approach, less flaps means a flatter approach. You want less flaps if the weather is very windy or turbulent, or more if you're trying to land on a short field. Half flaps was average for the Piper Warrior.

We were soon descending into sheer whiteness, as though the plane had been dropped into a bag of flour. The altimeter showed that we were less than a hundred feet up—leaving no margin for error. I glanced out the window and saw that the ground below us was looming quickly. The snow was so bad that I still couldn't see anything around the plane. Nonetheless I was afraid that I was still short of the runway. I prepared to open the throttle to maintain an altitude of about fifty feet, to make sure I didn't crash into the ground.

Just then, we broke out of the snow over the airport boundary. The problem was that we were about a hundred feet left of the runway. My heart in my throat, I made a sharp S-shaped turn, putting the plane down in a near perfect "greaser" landing—that is, a landing as smooth as grease, which is considered ideal.

As the plane rolled to a stop, I sat back in my seat. I expected both gratitude from Sherri and Paul, as well as a pretty exciting bit of documentary film-making.

It was not to be. "Oh my God, Dad, that was *AWFUL!*" Sherri wailed.

Paul, sitting in the back, looked as though he was waking from a coma. I asked him, "How did it look on camera?" He looked at me sheepishly, and responded, "I was too scared to shoot any film."

Apparently it was several years before he could be persuaded to step into any plane, even a commercial airliner.

xxxxxx

Perhaps the documentary film incident should have taught me a lesson about flying a malfunctioning plane. But it didn't. Several months later, my daughter Kimm asked if I would take her and her current boyfriend out for a Sunday sightseeing spin out of the Danbury Airport. I jumped at the chance, and rented another Piper Warrior from the FBO at Danbury.

I did a thorough preflight check, and turned up no problems. Everything—including the gyro that had let me down in January—seemed perfectly fine. We clambered into the plane, and I fired up the engine. The first sign of trouble came when I tried to hail the tower to get taxi clearance. I called, "Ground Control, this is Piper 4377P. Requesting clearance for taxi." Three or four times I tried, with no response. I attributed this to the heavy traffic; there were several planes seeking clearance at the same time. And after a few tries, the tower *did* give me permission to enter the taxiway, so I assumed everything was fine.

I waited in line on the taxiway for a few minutes. In the meantime, Kimm was clearly proud of her Dad's flying abilities, bragging about them to her boyfriend, who'd never been in a small plane before.

By the time I was number one for takeoff, there were about six planes waiting in line behind me on either side of the runway. Again, I had great difficulty getting in contact with the tower. Prudence would have told me to pull out of line and go back for radio repair. But Prudence wasn't in the plane that day. And anyway, pulling out of line was physically difficult, and I knew Kimm and her boyfriend would be disappointed.

So we took off and toured around the Connecticut shoreline. It was a blissfully beautiful, clear day, and a glorious ride. As I pointed out landmarks to Kimm and her boyfriend, I had nagging doubts about whether I'd be able to contact tower approach on our return. And, in fact, my fears were borne out. As we approached the field over the local VOR (a radio beacon used for navigation), I tried several times to get through. "Danbury Tower, this is Piper 4377P, requesting clearance to land." No response.

The galling thing was that I could clearly make out the conversations between the tower and other aircraft in the area. The problem obviously involved the microphone or some other transmission element of the electronics. But whether it was a rational problem or a gremlin in the works, it meant that I couldn't land at Danbury Airport. Danbury is a controlled runway, meaning that you have to have tower permission to land or take off. I was out of luck.

As it turns out, I knew about an uncontrolled field that was only a ten-minute flight away, in Oxford, Connecticut. I decided to land there, with the idea in mind of either getting the radio fixed and flying back to Danbury, or at the very least letting Danbury know about the problem and having them come and get the plane.

There weren't any mechanics available at Oxford. So I called the Danbury FBO and explained the problem. They refused to come and get the plane, and so I offered, "How about if you send out a plane to lead us back in?" Also a nonstarter. Instead, they insisted, "Fly the plane back. We'll contact the Danbury tower and tell them you're coming. Get into the circuit, and as you approach the field, put on your landing lights and the tower will signal you back with flares, and give you landing instructions."

This sounded easy enough. I got into the circuit without any difficulty. But as it turns out, there was another Piper Warrior, identical to mine in all but the identification number, immediately behind me in the circuit. I could clearly make out the conversation between the tower and *that* Piper.

It soon became clear that while the tower was talking to this other fellow, whose call letters were something like Piper RC938, they were looking at my plane, Piper 4377P. They didn't realize that there were two identical planes in the circuit, and neither apparently did the other pilot. I couldn't communicate with them to alert them to their mistake, and even though I was in the circuit, I certainly couldn't just land without the tower's OK; I'd lose my license. The only plan I could formulate was to duck out of the circuit and "buzz" the control tower, just like in the movie *Top Gun*.

When I started dropping out of the circuit, the tower noticed immediately. They began saying things like, "Piper RC938, you're coming in a little low," even though it was us—preparing to buzz the tower—and not the other chap who was low. Glancing back, I could see that he was in perfectly proper landing sequence. He responded, huffily, "I am *not* a little low." This exchange grew more and more heated—"Piper

RC938! You're too low! Pull up!" "I am NOT! You're *hallucinating!*" and so on.

As I got closer to the tower—and clearly out of the landing sequence—we could make out people in the tower, bailing out of their chairs and diving for cover, clearly believing that they were at the mercy of a kamikaze pilot. As we passed the tower, I waggled my wings. After a stunned moment, I heard the tower again: "Piper RC938, sorry for the mix-up. The plane we were looking at is that one that has no radio." After that, the tower gave me instructions, telling me to waggle the wings to signal that I heard them. We landed without incident—or, at least, without *further* incident.

xxxxxx

John F. Kennedy Jr.'s death at the controls of his plane sent a shudder of recognition through every private pilot. We have all faced circumstances similar to the ones he faced—iffy conditions, and an event we feel we have to attend.

In most cases, we put off the flight only to find that we could have flown perfectly safely if we hadn't "wimped out."

But occasionally we all tempt fate. I've done it at least once.

In the 1970's, the American Chemical Society asked me to organize a three-day course on nucleation at Mohonk House in the Catskills, in upstate New York. The course was supposed to start on a Monday. Getting to Mohonk from Cleveland was a bit of a problem. My options were to fly to New York commercially and drive for quite a few hours, or fly my own rented Cessna 152 to an airfield five miles away and hop into a cab for the rest of the trip.

This would be my single longest solo flight. I'd rack up eight hours of logged flying time. And more importantly it would be a macho trip that I could brag to my Mohonk class about. So naturally I decided to fly myself there.

On the day in question, I was thrilled to find that the Flight Service Station (FSS) reported good visibility at my start point, Chagrin Falls, Ohio, as well as at my destination, which was Kobelt Airport, near Poughkeepsie, New York. In between there were VFR (visual flight rules) conditions. (You might remember that this was exactly the situation JFK Jr. found himself in on the night of his fateful flight; he was told that there were VFR conditions between New Jersey, his takeoff point, and Martha's Vineyard, his destination.)

I filed a flight plan at the airport and took off. There were light gray cumulus clouds at about four thousand feet, but that didn't concern me since my trip would keep me at around 2,500 feet.

However, as I headed east the clouds got thicker and lower. As I crossed the Pennsylvania border, I did what every VFR private pilot swears he will not do. I climbed above the clouds, which were quite thin to 5,500 feet. Now, the sensible thing would have been to turn back and either wait out the cloud cover, or find another way to get to Mohonk.

But as any private pilot will tell you, the hardest maneuver in a private plane is a 180-degree turn; that is, a return to base. My plan instead hinged on the fact that I could see holes in the clouds where the ground was visible. As long as I could see the ground at least for a moment every minute or so, I would continue. If I couldn't, I'd turn back. Landing in Pennsylvania wasn't much of an option, since it is heavily forested and no possible landing sites exist for miles and miles.

After about three hours of flying, my intended refueling stop appeared magically through a break in the clouds. I landed, and as the Cessna was being refueled I noticed that there had clearly been a lot of rain at the airfield. The fuel attendant commented, "We haven't seen much traffic today, the weather's so rotten."

I was a little unnerved by that statement, but quickly rationalized that the FSS had assured me that my destination was clear. They hadn't used the dreaded words "VFR not recommended." So I soldiered on.

I was only about an hour from Kobelt as I took off, heading east. The clouds were clearly much thicker, and I assumed it was raining below cloud level. My rule about needing to see the land every one minute yielded to a two minute wait. Then five minutes. Then I stopped counting.

I reasoned that the "weather" was travelling east with me, and that I had to outrun it. I figured that with plenty of fuel on board, even if I couldn't land at Kobelt I'd have a couple of alternatives that were even further east.

My concern mounted as the clouds continued to thicken. I had conducted the entire trip relying on navigational instruments, but there wasn't any way I was qualified to pull off an instrument-only landing.

According to my calculations, I should now have been about ten miles from Kobelt. There was no break in the clouds anywhere in sight. Shortly I was nine miles from where I thought the airport ought to be. Then eight. Then seven. Finally at five miles, a miracle—the clouds parted and my destination was clearly in sight.

As soon as I landed, the tie-down crew at the Kobelt Airport asked me, "Did you come in from the East or the West?" When I responded "West," they commented, "You're the only plane who made it through VFR from the west today!"

I don't know whether it was admiration or contempt in their eyes. I like to think they were awestruck with my terrific flying skills.

But the fact is that if the weather had not opened a crack in the Pennsylvania clouds, I would have run out of gas and my body and airplane probably would have been found dangling from a tree several weeks later. Similarly, if the weather had reached Poughkeepsie before I did, I could easily have headed out over the Atlantic before finding a break in the clouds.

I swore I would never take those kinds of chances again. My resolve was soon put to the test. On the way home I headed west and could soon see a very gray sky with no definable horizon ahead. Soon there were thunderheads. I immediately put down at a tiny airfield, and stayed there all day while the rain and thunder passed overhead.

The next day dawned cloudless and beautiful, with visibility of fifty miles plus. As I climbed to 9,000 feet, the earth spread beautiful and serene below me. Having risked my life only two days earlier, I enjoyed the best flying of my life. But it was only luck that allowed me to live through my outward bound trip and savour it.

Chapter Twenty-Six

Exit the Ivory Tower

I have read about many artists who create their best work when they are miserable. The singer Billy Joel, the writer F. Scott Fitzgerald . . . there are many, many more. The 1970's proved that the same could be true for scientists.

My academic fortunes were at an all-time high between 1972 and 1976, when my personal life was a shambles. I had been promoted to full professor in 1971 at the age of 35, and was told I was probably the youngest person ever to achieve that rank at Case Western. I was managing one of the largest research projects at the University. I was routinely interviewed by newspapers and television stations. In fact, on the day my son Keir started his paper route, I was on the front page for snagging the largest research grant in the history of Case Western. I supervised truly outstanding students, and I believe we did some very creative work.

One project involved developing biocompatible materials, mainly for artificial heart parts. For that we had to figure out blood clotting systems and how they were initiated by interacting with plastics. This project laid some of the groundwork for modern-day medical implants. And it led my students to tape a picture of me to my office door, under which they wrote, "Beneath this gruff exterior there beats a heart of plastic."

We had all sorts of other programs as well. One involved attempting to figure out the molecular mechanism of aging and age-related diseases. This particular project helped my teenage daughter Kimm avoid wasting money at the cosmetics counter. I assured her that there was no face cream that could penetrate the skin, and that anyone who told her otherwise was putting her on. Whenever she went to department stores in search of makeup, she'd inevitably get a pitch for an expensive face cream as well, with all kinds of promises as to what these elixirs did at a "molecular level." "My father does medical research," she'd respond. "And he says you're full of it."

Probably the most commercially relevant program I directed involved calculating the potential for new pharmaceuticals and developing new drug delivery systems for oral insulin and anti-cancer agents. We also invented bioplastics and synthesized molecules that resembled fibrous proteins in the body, so that we could work out the structure of the real proteins.

In everything we did, we hoped that the fruits of our research would help people all over the world lead better, healthier lives. It was heady stuff indeed.

So things at the lab were cooking very nicely. My home life, on the other hand, was the antithesis of my soaring career. My first wife, Jasmin, had died in 1970. My second marriage, to Nancy, in 1972, lasted but a few torturous months. For several years after that I dated around, and my romantic life proved no more successful than my marriages had been. I assumed that there were women around who were not gold diggers or psychologically disturbed, but I clearly was not looking in the right places to find them.

Then, in 1977, I met E.J. on a flight to Toronto, and everything changed overnight. After our first date, I told my kids, "She can't be this wonderful. There must be something wrong with her." E.J. was fun-loving, caring, kind, and beyond all that, quite a dish. I was madly in love with her from the moment we met, and was over the moon when she agreed to marry me, six weeks after our first date.

Our marriage gave me a happy home life at last. Our blended family, with Kimm and Keir and E.J.'s two little girls, Kristin and Sherri, was a delight. Maybe a satisfying personal life gave me less creative drive at work. Or perhaps the stars or fate or luck turned against me professionally. But whatever the cause, my creative productivity quickly waned, and my ability to maintain a million-dollar-plus government grant program declined. By late 1978, I was ready for a new challenge, and I decided it was time to jump out of the academic nest.

I wanted to try my hand at business.

I thought that, after all, going into business wouldn't be a dramatic transition for me. I had for some time run a little business of my own, the Biopolymer Corporation, renting facilities from Case Western and paying post-doc and grad students to work for me in their spare time.

Biopolymers are sugars, or bits of things like protein which are hooked together in a long string to make one big molecule out of a lot of little molecules. We sold biopolymers of all kinds, those made of sugars and extracted from placenta or the combs of roosters, DNA extracted from cells or synthesized, and those made from amino acids found in proteins, which are called polypeptides. These were used mainly for research in molecular biology and possible therapeutic molecules. Many of these products sold for $10,000 or more for just one gram, far outstripping the value of gold. We had many more orders than we could fill.

One particular hot seller was a substance called hyaluronic acid, a white powder which went for a couple of hundred dollars a gram. I couldn't figure out why it was so popular, but I routinely had shady-looking characters visiting my lab who wanted to buy ten grams or more at a pop.

It was several months before I found out what was going on. It turns out that racehorses develop osteoarthritis of the knee joints from the constant pounding of running races. This makes many of them hobble to a halt at an early age, after a great deal of money has been invested in them. Racehorse owners would do anything to get one or two extra races out of a good horse, and it was routine for them to inject oils—any kinds of oils, including motor oils—into the horses' knees to "lubricate" the joints.

It seems that they somehow figured out that hyaluronic acid is the natural ingredient in joint lubricant, which is called synovial fluid. These racehorse owners were dissolving my powder to form an oil that was being injected into the horses. I don't know if it worked, but I *did* know that the practice was strictly illegal at the time.

Ironically enough, a Scandinavian company called Pharmacia now legally sells a preparation similar to mine. It's used for both horses and humans whose joints have deteriorated as a result of osteoarthritis.

With the Biopolymer Corporation, supervising my lab and consulting work all under my belt, how difficult could it be to make the move full-time into the business world?

I was shortly to discover just how dangerous naïveté can be.

By 1980, I had narrowed my options down to three. I had interviewed for the directorship of research and development for Abbott Laboratories, in their Chicago facility. It was unquestionably a prestigious job with a fabulous salary, but the building was drab and the thought of nine-to-five hours didn't appeal to me.

The second option was the diametric opposite of the Abbott job. I was offered the President and CEO position at a start-up biotech company called Valcor. The company was built on one of my own inventions for controlled drug release from natural body polysaccharides (polymerized sugars). Along with my assistants, grad student Randy Sparer and post-doc Nnoche Ekwuribe, I believed we had the foundation for a premier company. Unfortunately, the more I got to know the financial backers for the company, the less I wanted to work with them. Randy felt the same way. They were real sleazebags. Their payments for our research at Case started to become sporadic, and I quickly realized that I'd best consider other opportunities, despite the promise of the basic research.

The third opportunity was the one I eventually took. A publicly-traded company named University Patents wanted me to be President and CEO of a new subsidiary they were starting up, called University Genetics. UPAT had been in the university-based technology development business for nearly ten years, and their idea was to build on the hot new biotech/genetic engineering revolution by spinning out all of their biotech patent portfolio into University Genetics. Then UGEN would grow by funding university-based research and patenting the results.

The actual strategy for UGEN was not entirely clear, but I convinced myself that that was not a problem. I was blinded by the advantages I saw in joining UGEN. For one, I could leverage the business expertise of the UPAT folks. I understood the technology, and the idea of having an inside glimpse of cutting-edge research all over the country was a big thrill. And perhaps most importantly, at least from a business perspective, I could rely on UPAT to help me raise the necessary capital to get UGEN going.

UPAT's first idea was for me to take a two-year sabbatical from Case to run UGEN. I assumed that they weren't sure I could run a company, and were hedging their bets with a two-year contract. And for me, if UGEN didn't make it in the two years, I could go back to the tenured safety of my professorship at Case.

That idea was squelched when Case refused to grant my sabbatical request. "Commercial endeavor is not a suitable activity for academic leave," they answered. They countered with a one-year sabbatical, if I would sign a document promising that I would return after the year was up. I'm not sure how cutting a year from the sabbatical made a "commercial endeavor" any more suitable, but I felt that I couldn't in good faith make such a promise.

Instead, I made what was nearly the most disastrous step of my professional life. I resigned my academic position.

By June of 1981, everything was set. My family was packed up and ready to move to Connecticut. We had chosen a new contemporary home in Weston, making the down payment by cashing out my retirement plan at Case. The investment bankers Allen & Company were promoting the UGEN prospectus, and the future for the company seemed bright. Everyone told me to anticipate proceeds from the private offering in the $27 million range, plenty to get UGEN off the ground.

On July 15, 1981, a happy caravan of E.J. and me, our four kids, two dogs, and all our belongings made its way to Connecticut.

Two months later, I was essentially jobless.

The offering had collapsed. What I hadn't realized was that the stock market had gone south, and biotech stocks, which had been the hottest sector of the market, suddenly went cold. Despite this, the bankers had been telling us every day, "Everything is fine" and "Things are going very well." It wasn't until later that I learned that even after the bankers have thrown up their hands and gone on vacation, they go on saying these things until the day that a financing actually collapses.

UPAT offered to pay me a reduced salary for one year, "and then you're on your own."

I was a wreck. I felt naïve and stupid, an utter failure before my business career ever got off the ground. Here we were in a brand new house which had blown our entire savings, having uprooted our family to move five hundred miles away. And I had been too foolhardy to think about "option B"!

I was paralyzed with failure. It was undoubtedly the darkest moment of my professional life. Without E.J.'s support, I have no idea what would have happened. Her sunny disposition was the reflection of a cast-iron optimism, and the utter determination to make *sure* that things worked out for the best. Rather than let me mope, she prodded me to think of other alternatives for UGEN. I talked to Bill Miles, President of UPAT, and together we hatched a revised business plan, with the idea in mind that we would hawk the UGEN prospectus to any wealthy potential backers who would hear us out.

Ultimately we raised a $3 million research and development financing in December of 1981. That got UGEN off the ground.

The five and a half years that I spent running UGEN were both exhilarating and frustrating. I quickly found that my original plan—to fund university-based research, and exploit the results commercially—wasn't going to support UGEN. It was a classic armchair situation where what seems foolproof in theory falls apart in practice. While I, and everybody else associated with UGEN, thought that university researchers would be thrilled at the prospect of commercial success, what we found instead was that their primary goal was to continue receiving funding for the research. We pleaded, cajoled, begged, and threatened, but squeezing commercial-grade results from university labs was excruciatingly difficult.

My "Plan B" was to form small biotech businesses based on university-originated technology and run them as subsidiaries. This strategy worked quite well, at least for a while.

One of the companies we formed revolved around an automated machine for examining chromosomes, based on technology invented at the George Washington University. We offered the university 5% of the equity in the new company in return for the technology. At the time, GW—in common with most universities—had no idea what to do with stock. Instead, they asked, "Can we have a 5% royalty on sales instead?" Absolutely. And we proceeded with that deal in hand.

We spent $50,000 forming the company and patenting the technology. We found investors to put up $180,000 for half of the company. We hired management, brought in additional technology, and took the company public within three years. We wound up with $2 million in cash for our $50,000 investment, not a bad return.

As a result of this experience, I'm pretty sure GWU revised their attitude toward accepting stock. The instrument we licensed from them never wound up being put into production so they didn't receive any royalties after all. If they'd taken the equity, they'd have received about a quarter of a million dollars. Ouch!

Along with starting up small biotech businesses, we also continued to license biological technology from universities to large companies or start-ups. The most successful of these licenses involved a DNA synthesizer invented at the University of Colorado.

We licensed the technology to a start-up company called Applied Biosystems in the early 1980's. The timing is important, because it had only just become possible to offer exclusive licenses for government-funded research (which is what the Colorado research had been). The Dole-Bayh Act allowed this kind of exclusive licensing, and I was particularly familiar with the Act, because I'd worked on elements of it in Washington in the late 1970's.

The only fly in the ointment was that a large company, Beckman Instruments, objected to the license being given to a start-up with only one employee (the now famed Sam Eletr). Because they kicked up such a stink we were forced to grant them a co-license, along with Applied

Biosystems. So both companies were free to manufacture and sell the DNA synthesizer.

Fast forward three years. Applied Biosystems had the instrument on the market. Beckman was still digging a hole for the building where they were going to start the work. Applied Biosystems went on to become one of the biggest biotech success stories of the 80's and 90's. It was eventually sold to Perkin Elmer for several billion dollars. After that, Perkin Elmer sold off all its other businesses and became Perkin Elmer Biosystems, a Fortune 500 company. It tickled me no end that rinky-dink little Applied Biosystems had beaten Beckman, a huge bureaucratic company, to the punch. People who think that large companies by dint of their vast resources have a huge advantage are wrong, wrong, wrong. I've seen many situations like Applied Biosystems, where a small company has the real advantage, because it can move and execute so much more quickly than a huge one.

The licensing division remained a money maker for UGEN throughout my tenure. By the time I left the company, there were three other operating divisions as well. One was Applied Animal Genetics, a California-based company which used technologies like embryo transfer and embryo splitting to build superior herds of Holstein cattle worldwide. It was a pretty remarkable company. With AAG's technology, you could carry a hundred frozen embryos under an airplane seat to Indonesia (what a good time explaining *that* to the carry-on baggage security folks!), implant the embryos in inferior local animals, and produce five dozen top-grade Holsteins.

Another of UGEN's operating divisions was Agrogene, based in Florida. It imported rare plants and cloned them. Basically Agrogene did for plants what AAG did for Holsteins. You could take one rare and expensive plant, clone it, and raise a whole greenhouse full of identical plants. At least, that's how it was *supposed* to work. Actually *growing* plants is subject to the same vicissitudes of weather that have dogged crops since the beginning of time.

The other UGEN operating division was American Diagnostic Sales. It marketed cow and horse pregnancy kits. Through ADS, everybody at the office learned more than they ever wanted to know about the courtship rituals of farm animals.

All told, the four operating divisions at UGEN employed about a hundred fifty people. While I had always managed people in an academic setting, I found it unexpectedly challenging in business. UGEN attracted all kinds of characters. There was one scientist we hired at one of the subsidiaries who almost immediately proved to be a problem. We had brought him on in the first place because he had come up with an invention that he assured us was his own. That turned out not to be the case, and he was apparently a thief of both intellectual and tangible property. We even found that an optical microscope he had brought to work with him—which he claimed was his own—had a sticker on the bottom of it reading "Property of University of XYZ"—his previous employer!

Another "problem employee" cropped up at one of the subsidiaries. One day our in-house lawyer, Tom Monahan, got a call from the bookkeeper at this particular subsidiary. "We've got a situation here," the bookkeeper said. "I asked Fred [not his real name] for his Social Security number, for tax purposes. He told me, 'Oh, I don't have to pay taxes. I'm a Congressional Medal of Honor winner.'" Apparently the bookkeeper followed up on this, by calling the appropriate authorities in Washington, D.C. Sure enough, Fred was not a Medal of Honor winner.

Tom decided to pull out Fred's résumé so he could do a little sleuthing and see if anything else was suspicious. Fred claimed that he had an L.L.B. degree from the University of Central South Carolina. Tom asked my daughter Kimm—then a newly minted law school graduate herself—if she'd ever heard of an L.L.B. degree. "Sure," she responded. "It's an old-fashioned name for a law degree. They still call it that in England. But that's not your problem," she went on. "Your problem is that there *is* no law school at the University of Central South Carolina."

It turns out that virtually everything on Fred's résumé was a total fabrication. When Tom called the bookkeeper at the subsidiary and told him to confront Fred, the bookkeeper responded, "Are you kidding? It's not just that he's my boss. He's about six foot five and he packs heat, a .357 Magnum. You want to confront him, *you* come down here and take him on. Not me!"

UGEN was even more of a challenge because it was chronically underfunded. Starting with $3 million instead of the $27 million stock offering I'd anticipated meant that I was constantly sweating our financial prospects in the near term. Meeting payroll was a frequent worry. The best story I know about payroll problems is one that happened to a friend of mine, Dr. Sam Waksal, who was president of a small pharmaceuticals company called ImClone Pharmaceuticals.

ImClone, like UGEN, was a public biotech company with a seemingly permanent cash shortage. Sam was therefore delighted when he was able to negotiate an $11 million contract with a well-known German pharmaceuticals company that I'll call Munich Pharmaceuticals. Eleven million dollars was just the shot in the arm ImClone needed, since Sam had barely enough cash in ImClone's checking account to meet one more payroll. Two weeks without a cash infusion, and the jig would be up.

So he rushed off to Germany to sign the final documents with Munich. When he got there, the Munich folks hemmed and hawed. They told him that they didn't think the deal was worth $11 million after all, but more like $5 million. Decimated on the inside yet cool on the surface, Sam negotiated with them for an hour, to no avail. Finally, Sam said, "Look. I am going to the bathroom. When I get back, I want the $11 million deal or I am going down the street to Gerhardt Pharmaceuticals [not its real name], to sell them the contract."

He in fact had no contacts at Gerhardt and no chance at all of selling them the contract. It was Munich or nothing. So he went to the bathroom, and as he tells it, "I got down on my knees and prayed, for the first time in my life." Composing himself, he got back up and returned to the conference room. The Munich spokesman said, "All right. We'll do the deal for $11 million."

This kind of brinksmanship is the stock in trade of the captains of small industry!

Financing and employee issues aside, there was a lot to love about running UGEN. I appeared on all kinds of television and radio shows and got interviewed all the time by the print media. Biotech was a relatively new thing, and I seemed to develop a reputation for being able to translate it for the viewing, listening, and reading public. As I've always been a bit star-struck, this was a great thrill for me.

I also got a kick out of dealing with shareholders. Biotech, with its sci-fi aspects, generated all kinds of interesting questions and comments. I remember in particular one lovely little old lady, who called me to tell me that her pet cat had died, and she had put Muffy in her freezer "until you figure out a way to clone her."

XXXXXX

As time went on, it was clear that UGEN needed additional rounds of financing in order to grow. This proved to be the company's undoing, but not for any conventional reason. Instead, it was UPAT's unique voting position that doomed UGEN. UPAT insisted that their shares in UGEN had to be of a different class than every other shareholder. The special UPAT shares would have 4-to-1 voting rights. In essence this prevented any financings, because there weren't sufficient "1-vote" shares to achieve a voting majority.

When it dawned on me that UPAT had a choke-hold on UGEN, I realized it was time to move on. The publicity I had received through UGEN made it relatively easy to jump into a new career—venture capital.

So I resigned from UGEN in 1986, leaving my friend and fellow Brit Randal Charlton as President and CEO of UGEN. Randal is one of the nicest guys on Earth. Beneath the pleasant exterior, though, is a proud competitor.

Randal went on a fishing trip once with Keir and me in the wilds of Montana. We rafted down some churning waterways and caught fish at the calm spots in between.

One day, everybody on the raft caught fish, except Randal. As we approached some fairly violent rapids, our guide shouted to everybody to get their fishing rods back in the boat and hang on. Randal refused to stop fishing; he was *determined* to catch a fish, even if it meant risking his neck. I quickly forgot about Randal as we entered the rapids, with water thrashing around us as though we were being shaken in a washtub.

After several minutes in the maelstrom, we emerged to calm waters on the other side. I looked around the boat, and there was Randal, standing up, grinning ear to ear. He held up his fishing line, and there, on the end of it, was a fish.

An eternal optimist, Randal didn't share my doom-and-gloom prediction for UGEN. In fact, he laid out a strategy for UGEN that had never dawned on me. He decided to sell off all of UGEN's subsidiaries except its plant genetics division in Florida. It was a very sexy business. The facility could bring in an exotic plant one year, and by the following year, produce 100,000 perfect clones of it.

This, to Randal, was marketing heaven. He foresaw selling the plants to chain stores for $5 to 10 a pop. And I'm sure he could have. But the problem he faced was familiar to any farmer: Acts of God get in the way. In the case of the cloned plants, there were two years in a row when a virus or pathogen wiped out the entire stock just prior to shipping. The losses were staggering, and led to UGEN's sad demise.

Ironically enough, the big winner in the UGEN saga was UPAT. UPAT's total costs for the entire endeavour amounted to about $60,000. Five years after this initial investment, they distributed their then-public UGEN stock for $3.6 million, making an astronomical sixty-times profit in five years. It remains one of the biggest returns in the history of the biotech industry.

Chapter Twenty-Seven

Venturing into Venture Capital

If there is one theme that connects every career change I've ever made, it's this: I've never paid attention to where the money comes from.

When I first dreamed of becoming a professor, I thought about how great it would be to do research in an atmosphere of unfettered intellectual stimulation. I loved the research, but I hated the grant writing, the money that greased the wheels of academic endeavour.

As an academic, I thought that all I had to do was come up with a great idea, patent it, and then businessmen would beat a path to my door and pay me zillions of dollars to license my patent. I'd be set for life. It didn't happen.

When I got to UGEN, I quickly learned that only one in ten patents is licensable at all. Most patents just aren't commercially feasible. And of the very few that *are,* only one in five generate more than $20,000 a year in royalties for the inventor.

Instead, what took up a lion's share of my time at UGEN was going on metaphorically bloodied hands and knees to venture capitalists, looking for financial backing. It was incredibly frustrating, knowing that we had fantastic technologies which I was certain would generate huge sums of money, if only we could convince venture capitalists—none of whom understood a whit about science—that we were sitting on a gold mine.

So I started to think—Aha! Venture capital is surely heaven on earth. As a venture capitalist, everybody—inventors, businessmen, creators and dreamers—all of which I'd been—would come to me, begging for money. I'd get to hand it out. What could be easier?

Of course, what wasn't obvious from the outside was that venture capital money itself doesn't zip out of the office copier, ready to be distributed. A lot of time in venture capital gets chewed up in raising a venture capital fund. Which involves—no surprise here—going on metaphysically bloodied hands and knees to investors: Pension funds, foundations, corporations, and the like.

I have learned better now. I have *no* desire to join a pension fund, foundation, or corporation.

But in fact venture capital is great fun. Aside from its fundraising aspects, that is. So I found ultimately that the grass is greener here.

In a nutshell, this is how venture capital works. Partners in venture capital firms go out and talk to all kinds of institutions and wealthy individuals, convincing them to invest in the fund. Venture capital funds these days vary in size from $10 million all the way to $10 billion, depending on the specialty. In the life sciences, there are probably no more than a dozen or so that are bigger than $100 million. The typical deal is that the venture fund will last ten years, and then the investors will be paid off. The ultimate goal is to take all of the invested companies public, or sell them to appropriate buyers, to make it easy to cash out at the close of the fund.

Once the fund is raised, the fun really begins. The partners invest the fund's money in businesses, from start-ups looking for seed money to existing companies looking for "mezzanine" financing. Once the money is invested, the partners often sit on each investee's board of directors to see that things are running properly.

XXXXXX

My own entrance into venture capital came about by chance, as do most career changes.

Because I had a fairly broad knowledge of biotechnology and had co-authored a number of books on the subject, I was in some demand as a consultant to venture capital funds. While they knew everything there was to know about business—and I'm not sure how much that is!—it's hard to invest in biotech companies without understanding the "biotech" part. At the time, almost no fund in the world had a scientist in-house capable of understanding the molecular biology associated with biotech (now, virtually every fund has at least one Ph.D. on staff). One of my consulting clients was Oxford Partners. After I'd been consulting with them for a little while, they invited me to join them.

Becoming a venture capitalist was a real trial by fire. I was joining an established venture capital fund as a true outsider. That had at least one benefit; I didn't have to start out by raising a fund. And thank God for that! When I did later raise my first fund, I found that the problems were exactly the same as they were when I tried to get grant money as an assistant professor. You need performance to raise capital and you need capital to fuel performance. The major venture capital investors want to see funds which have three partners who have worked together for ten years, have had annual net returns of fifty percent, and they want all of the partners to be under fifty years of age.

There is not one venture capital firm in the country that fits all of these criteria.

My favourite story about wooing investors has nothing to do, at least not directly, with wooing investors.

In 1991, the old Oxford fund—the one I'd joined as a partner—was winding down, and I was raising my own fund, focusing exclusively on biotech. I went to London by myself to make a pitch to the Wellcome Foundation. Their London office was close to the Euston railway station. I arrived at Euston early for my 1 P.M. meeting, and I was famished. I figured I would duck into Euston station, pick up a meat pie, and eat it before the meeting.

The odours from the meat pie shop were scrumptious. My mouth watered as I stood in line, waiting to place my order. I paid for it, and then shuffled down the counter to wait for my meat pie.

Just as the counter clerk was sliding my pie onto the counter, a police officer rushed in and yelled, "Everybody out! IRA bomb threat!"

The crowd immediately dispersed. I made a quick lunge for my pastry, determined to grab it and run out of the station. The police officer, standing nearby, grabbed my arm and demanded, "What are you doing?" I protested that I was just trying to grab what was rightfully mine. He looked at me as though I'd lost my mind and asked, "Do you want your meat pie—or your life?"

He wrenched me away from the counter before I had the chance to decide in favour of the meat pie. For all I know, the pie is still there. I left, lacking both the meat pie and the money I'd paid for it. But I did have a great opening line when I got to Wellcome.

They didn't find it sufficiently entertaining to actually give me any money, mind you.

XXXXXX

When I jumped into venture capital, I felt—as I had done many times in my life—like a complete outsider. Virtually everyone in venture capital has an MBA. Those without an MBA have a business undergraduate degree. As far as I know, I am the only former tenured professor in the venture capital industry. As a result, I brought a very different sensibility to the table. I knew literally nothing about any other kind of investment Oxford made. My only focus was on its biotech portfolio.

From the beginning venture capital was radically different from anything else I'd experienced. For a start, there were the checks I was writing. In everything else I've ever done, a million dollars is a lot of money. A *lot* of money. I'd never conceived of actually writing checks for seven-figure sums. But in venture capital, I routinely write checks for $2 million or more, to get young companies moving. (In the world of big business that's still chicken feed. I have a friend who used to be president of the World Bank. He told me that he was once staying at a hotel and had to make a wire transfer that very day. He wouldn't have a chance to get back to his office. So he went to the front desk, and when he told the hotel clerk he had to make a wire transfer, the clerk casually asked, "How much?" and he responded, "One billion dollars." "The guy almost had a heart attack," my friend laughingly told me.)

I was also dazzled by the impact venture capitalists can have. We get to see and evaluate cutting-edge technology, ideas that, when they come to fruition, may help mankind decades into the future. With my scientific background, I have an overpowering desire to see new technology get commercialized. I dream of seeing new medical devices brought to market, and drugs developed for new and currently incurable diseases.

It excites me to think about the next fifty years. Human organs and diseases will be modeled by computer. Critical disease targets and biochemical pathways will be identified. Candidate therapeutics and their toxicology will all be predicted by computer. The random "hit and miss" methodology of the twentieth century will be turned into the science of drug discovery, which will make new pharmaceuticals available quicker, cheaper and more effective at preventing or treating diseases.

XXXXXX

After venture capitalists finance exciting new companies we often contribute the operating experience and financial tools to make these dreams real. There's *no* thrill like it. Of course, incubating new technologies is one thing; but generating revenues for investors is an entirely different challenge, since they tend to have a shorter horizon than the five years typically required to grow a biotech company. It's a dilemma that we deal with every day.

Before I joined Oxford, I thought venture capitalists would be cowboys, real risk-takers. After all, they were investing in ideas! In fact, venture capitalists and entrepreneurs are very different. I quickly found that entrepreneurs often believe that there is no way they can fail. They'll mortgage their house and their children to pursue their dreams. Most venture capitalists are just the opposite. They believe that if anything can go wrong, it will. That philosophy makes them very conservative in placing their bets.

It's impossible to be in a business like venture capital, involving huge amounts of risk, without coming up with *some* bellwether for deciding whether or not to take the plunge with any particular company. Many people have spent a great deal of time trying to quantify what makes a company successful. The answer? There is no answer.

One of my friends in the industry, Bill Boeger of Quest Ventures, once told me that he and his partners went through dozens of failed companies that they'd invested in over many years. Looking for common weaknesses, they had expected that management, technology, location, availability of finance or other well-known factors would be the common failing.

Much to their surprise, they found that the only parameter in common in their failures was that in each case, four or more "blue ribbon" investors were involved!

At first, we laughed and felt that this was just a curious coincidence. But as I thought about it afterwards, I realized that when large groups of investors are involved in a company, they each tend to assume that the others all know what they're doing, and so the ball seems to be dropped when in fact nobody actually held it in the first place.

Of course, *some* investment philosophies make sense. One bit of venture capital dogma is that one should not be the only venture investor in a company. The reasoning lies in the downside risk. If the company fails, everyone laughs behind your back and claims that they didn't invest because they knew the company would fail. On the other hand, if several venture capitalists invest in a failed company, they blame the management.

Perhaps the ultimate example of the fatal "one investor" syndrome is that of a northeastern university which formed a small venture capital group to invest in a technology invented by one of their professors. They believed this invention would be a home run in curing cancer.

Being greenhorn venture capitalists, they fell into a very common trap: they overlooked the unrelenting cash drain of clinically-oriented biotech companies. After spending almost $80 million of university money under the direction of the university's president, the company was taken public. Its value steadily declined until it was sold out of bankruptcy for less than $20 million. The university had lost more than $50 million in this one investment alone.

xxxxxx

One common pitfall of venture investing is a variation on the old chestnut that if you owe the bank $10,000 and can't pay it, the bank owns you. But if you owe the bank $10 million and can't pay it, you own the bank. (Some people cynically refer to this as the Trump Phenomenon.)

The pitfall is called the "black hole syndrome," and it covers this situation. After you've invested a couple of million dollars in a company, it's very hard to label it a failure. The temptation is to say instead, "Well, with another million or so the company will make it." The biggest "black hole" loss in history is reputed to be Exxon's venture investment in amorphous semi-conductors. They reportedly lost upwards of $300 million on it.

The most tragic losses in venture capital are the ones that are the result of pure fraud. No amount of due diligence in the world will guarantee that you'll never get taken by a crook. One famous company had rapidly growing computer sales, but its cash flow didn't jibe with the amount of product it was shipping out. It turns out that the company was shipping bricks to warehouses, and counting them as computers!

Every venture fund has a company now and then that "cooks the books." One of the companies from the first Oxford fund had executives who are now in prison for doing just that.

One of the saddest situations I ever came across was a company we invested in that had created what seemed to be a very promising pharmaceutical. The company—I'll call it XYZ—had been started by another venture fund with whom we'd had very successful dealings in the past.

XYZ's president was a very charming scientist who had flawless credentials. I'll call him Joe Chen (not his real name). He'd been very successful in R&D at a large pharmaceutical company before starting XYZ. As is true of many presidents of small companies, Chen was under tremendous pressure to bring in additional financing and to entice a corporate partner to invest in XYZ. In presenting data to a potential partner, he "slightly modified" two of the slides he presented. The corporate partner invested, as did several other venture firms.

The forgery was discovered by the Director of Research and Development for XYZ, who could not duplicate the data or identify its origin. He notified the Board of Directors, a very brave thing to do since such a revelation would completely undermine everything on which the company was based.

All hell broke loose. Everyone demanded their money back and the company eventually collapsed. The biggest investor, the venture fund that had founded the company, lost $9 million on the deal.

I was confounded by what happened, and not least of all because I really liked Joe Chen. I sat down to talk with him about it, curious to find out why he'd taken such a risk. He was clearly suicidal. He told me, "Alan, it was the pressure. The success of XYZ depended on good data. My personal and family stature was at stake. The success of the investors—it was all depending on me."

His career was clearly over. But I like to think that by lending a sympathetic ear, I saved him from killing himself.

xxxxxx

One of the occupational hazards of venture capital is being sued for something you didn't do. It's the promise of quick bucks—even if they're unearned—that leads people to file unscrupulous lawsuits against venture capitalists. It happens all the time.

The biggest groundless lawsuit that I've ever been involved in concerned a company we invested heavily in during the late 1980's. I'll call it Sorbonne Pharmaceuticals. Within four years, we were able to take the company public at $15 a share, raising $70 million or so. The company's strength at that point was two drugs it had in preliminary human testing with the FDA. One treated a skin disorder, and the other was a cancer drug.

I felt that in both cases the drugs had been pushed prematurely into human testing so that the company could go public. My fears were only somewhat quelled by the fact that several people who'd worked for large pharmaceutical companies sat on the board of Sorbonne, and they all said that the quality of the drugs was such that their former employers would have gone ahead with them.

In fact, as time went on, the board divided into two factions. One other board member and I were a distinct minority, pushing for the company to slow down its spending and concentrate on doing more fundamental research. The majority believed that Sorbonne should goose up its expenditures on clinical trials, to push its two drugs—the ones I thought weren't particularly good—through the FDA.

Naturally, the majority had its way and Sorbonne's cash burn rate skyrocketed. I felt sure the company was headed off a cliff. Predictably enough, the skin disorder product failed in clinical tests. That led Sorbonne to lay off some of its staff, and made the stock drop like a stone—it went from $15 a share to $3 in short order.

With a little quick footwork I managed to get Sorbonne to acquire another company, one whose technology I thought was super. Sorbonne got the company in return for stock. Shortly afterwards, one of the biggest pharmaceutical companies in the world acquired Sorbonne for $100 million. The new owner immediately threw out Sorbonne's clinical program—the one I'd been opposed to in the first place—and grew the technology I'd arranged for Sorbonne to acquire. The purchase price for Sorbonne was $6 a share—double its market value.

Immediately after Sorbonne was sold, we received a cookie-cutter lawsuit from an east coast law firm. It began, "On behalf of shareholders of Sorbonne Pharmaceuticals too numerous to name . . ." The claim was essentially that we had sold the company too cheaply. This is a very easy thing to say—you can claim it of *any* sale. But in fact we knew that virtually every shareholder was delighted that we'd salvaged the company. This law firm had no case.

But it didn't matter. We were all put through extensive questioning by lawyers and chewed up mountains of time on what amounted to a nuisance suit. The result? The big pharmaceutical company that had bought Sorbonne wound up paying off the law firm to the tune of $1.5 million. I don't think the shareholders ever received a dime of it.

xxxxxx

My own investing philosophy is one I carried with me from my years as a research professor. I look for two things: either major new technologies that interface between two existing technologies, or a significant new step forward in a single technology.

Of course, because as venture capitalists we have money to hand out, we see thousands and thousands of people with ideas and inventions, and embryonic companies looking for backing. Some of the business plans we get are bizarre. I remember one plan that pressed the use of indium as a kind of wonder drug, the promised benefits ranging from softened ear wax to prolonged erections. The inventor had overlooked one slight problem with indium. It's poisonous.

Another business plan called for burials in space. (Of course, *that* one only seemed crazy until a company actually started launching people's ashes into the outer atmosphere—*Star Trek* creator Gene Roddenberry was one of its first "passengers.")

Other ideas we see are brilliant. Ironically, many of the most creative and potentially commercial new ideas we see don't come from academics, but from existing biotech companies. That makes sense in that an increasing percentage of national research dollars are spent in companies, not at universities. One advantage of having ideas come from companies is that they know what it means for something to be commercially viable. Many academics, clever as they are, have great ideas. But the ideas are often only theoretical, or the inventor has no idea how to commercialize it, and thus can't run a company based on the idea.

Biotech is a particularly challenging platform for investment, because it takes years and years to develop a product, and only the most successful companies generate traditional venture capital revenue by being bought by other companies or by going public. In the meantime, it's very difficult for venture capitalists and the public to know how to value unproven technology, figure out the strengths of management, determine commercial viability, and assess adequate financial resources. While venture capitalists are prepared to wait four or five years or even longer to see value in a company develop, the public isn't that patient. If the public doesn't see significant financial progress in six months, it's likely to sell. That makes biotech investing extremely stressful.

xxxxxx

The promise of untold riches leads many people to abandon their principles. There have been many technology related companies (none that I've backed, thank God) based on questionable, if not downright phony, science. My favourite story about this involves an Israeli professor who visited me several years ago. He claimed to have a new cure for a particular disease, and he wanted two million dollars or so to capitalize on it. As he talked, I seemed to remember that I had seen something similar to what he was proposing. On top of that I was pretty sure that the process he was describing wouldn't provide the cure he promised. Nonetheless, I wanted to be polite, so instead of confronting him directly, I asked, "What happens if your cure doesn't work?"

Without hesitation he replied, "That's easy. We'll take the company public, of course."

xxxxxx

Venture capitalists get their hands on all kinds of inside information about companies. We sit on the boards of directors of all of our portfolio companies, and that means that we know, at least in theory, exactly what a company is up to before that information reaches the public. This inside skinny creates all kinds of "insider trading" opportunities; that is, the chance to make illicit profits on non-public information.

For instance, let's say you sit on the board of Little Fish Corporation. In that role you learn that Big Fish Corporation is going to buy Little Fish for much more than its stock value. Anybody who buys stock before the buyout information becomes public is going to make a pot of money, since the stock will unquestionably go up in value when the sale is made public. But trading on that kind of information is strictly illegal. That's why my partners and I have strict rules against investing in any of our portfolio companies on a personal basis. This is a universal rule in venture capital.

Unfortunately, sometimes people have difficulty avoiding the temptation of insider trading. In fact, it happened to a former partner of mine—with tragic consequences.

In the late 1980's, Oxford invested in a company that I'll call Magnus Pharmaceuticals. I was on the board of directors. Magnus went public in the early 1990's. It subsequently began highly secret meetings with a large Japanese company that was interested in buying Magnus.

These meetings were so secret that the Board would meet in various non-company locations in downtown Washington on weekends. The CEO of the company, recognizing that there were several venture capitalists on the board, asked us not to inform our partners about the situation since it was so sensitive. We agreed. When I'd go to our weekly meetings at Oxford to review the progress of our portfolio companies, when we came to Magnus I would generally give a vague report that things were going well, and leave it at that.

Unfortunately, a former partner of mine, whom I'll call Fred Sellers, apparently liked what he heard about Magnus at these weekly Oxford meetings. Without knowing anything about the potential windfall from the possible buyout, he bought a chunk of Magnus shares for himself, his wife and his children—a cardinal no-no. Now if Magnus had just chugged along with its stock rising gradually, perhaps he'd never have been caught. But as it turns out the Japanese company soon bought Magnus for many times its book value.

The Securities and Exchange Commission combs through all records for evidence of insider trades, and the Sellers trades immediately popped up. The SEC was immediately all over him. They investigated him for three years, and questioned all of the Oxford partners over and over again. Sellers wound up being fined almost a million dollars for his insider trades.

Needless to say, Sellers is no longer in venture capital.

xxxxxx

Venture capital is an exciting business, and many people are attracted to it as a career. When I interview potential associates, I often ask them, "What makes you think you'll be a good venture capitalist?" They generally answer with comments about their background, interactive skills or even their understanding of technology or people. But my hands-down favourite answer to this question came from a young South African M.D. He had started several companies and seemed to have an ideal background for venture capital. He nonchalantly answered, "Oh, I have the perfect qualification. I have a very short attention span." We offered him a job in spite of that answer. He turned it down, opting for a position in the pharmaceutical industry, but alas, he didn't do well.

Maybe it was that short attention span.

xxxxxx

One of the occupational hazards of being a venture capitalist is that once people find out what you do, they suddenly have a business in which you just *have* to invest. Because I focus exclusively on biotech, I'm able to wriggle out of most of these uncomfortable situations. But sometimes it pays to say "Yes," no matter how diametrically opposed the idea is to your investment strategy.

I learned this lesson in 1994. We'd been out for a Mother's Day brunch for E.J. Halfway through the meal, I suddenly couldn't see anything through my left eye. Everything went black.

It turns out I had a detached retina. The doctors told me that these types of injuries are caused by age, or sports accidents, or too much sex. While I'd love to brag that it was one of the latter that caused my injury, it was probably advancing years that were to blame.

The surgery to correct the problem was remarkable. The surgeon had to inject relaxants into the muscles around my left eye, remove it, and literally sew it up from the back. The recovery was miserable, alleviated markedly by E.J.'s wonderful nursing. I had to lie on my side for ten days, keeping my eye as still as possible. Whenever I accidentally moved it—try keeping your eye looking directly forward sometime—it felt as though I had gravel grinding against the back of the eyeball.

Nonetheless the surgery was a success. A little while later I had to travel to Johns Hopkins to get minor corrective surgery performed on it. After preparing me in pre-op, probably with a sedating drug, the assisting surgeon remarked, "I understand you are a venture capitalist. You invest in biotechnology." I mumbled something incoherent, but he went on, "Well, I've discovered a gene involved in macular degeneration. Are you interested?"

As the sedative took effect, I managed to tell this savvy pitchman that if he did a good job on the surgery, I'd take a look at his business plan. I wasn't in much of a position to haggle.

The surgery went fine. But I never did get that business plan.

XXXXXX

When I joined Oxford in the late 1980's, with my background in academia I was certainly one of the few people in venture capital who knew anything about biotechnology. The industry was thickly populated with Harvard MBA's, all of whom were very intelligent, but when it came to biotechnology they were investing blind. The philosophy behind this was that "We invest in people, not technology." I was told over and over again that there were three criteria for an investment: Management, management, and management.

When it comes to biotechnology, this philosophy is bull, bull, bull. As I got my feet wet at Oxford and studied our portfolio companies, my attention was immediately drawn to the only biotech company. I was familiar with it, because as a consultant with Oxford I'd warned against investing in it.

My reasoning was that the founding scientist had refused to reveal details of how he had come up with a universal anti-allergy agent. My hunch as a scientist was that the reason for his secrecy was that the technology wouldn't work. Nonetheless, Oxford had backed this fellow. When I questioned this decision, I was told, "All the best venture firms are investors. It must be a good company."

With this hardly encouraging information, I started sitting in on board meetings of this particular company. I couldn't believe the misinformation that management was promoting. For instance, when they were asked why they hadn't co-ventured their anti-allergy compound with a pharmaceutical company—a logical move for a company in their position—they explained, "We want to keep all the upside profits for the company." When asked why the FDA kept turning down permission to market their drug, the answer came back: "The FDA has political motives for not wanting to see a small biotech company like us succeed." Incredulous, I glanced around the large round table, to see a dozen experienced venture capitalists nodding in agreement at this pile of baloney.

In fact I had done a little sleuthing of my own. I'd learned that the company had, in fact, approached every pharmaceutical company on earth, looking for a co-venturer. Their pitch was that the drug was effective and lasted in the body for a week. They'd been turned down universally because the pharmaceutical companies had all recognized that not only did the drug not work, but it disappeared from the body within a very few minutes through breakdown and excretion.

The company eventually persuaded a Japanese outfit that it had a valuable product, and all told the company swallowed up $40 million before it became common knowledge that the product was a bust. You'd think that with that kind of experience, the company would have had to declare bankruptcy. But it didn't. Instead, it used some of its money to buy a small company that sold nasal inhalers. It then persuaded an established investment bank that the anti-allergy drug was for real, and they took the company public.

Subsequently, the Japanese company—recognizing that it had made a huge mistake—took all of the research and development activities out of the company and renamed it. The anti-allergy compound was promptly dropped, and the company subsequently became successful with other products. Talk about creating a silk purse from a sow's ear!

xxxxxx

We've backed lots and lots of interesting companies. One of the ones funded out of the original Oxford Partnership was Martek. It produced specialty chemicals from algae—the scum you find on ponds. They managed to form a very interesting business producing food additives from genetically-modified algae. One obvious benefit of this business model is that you can get your raw materials for free—the market for pond scum is forever depressed. And the market for fermented algae is huge; there's pond scum in many of the things we eat.

Another of our "investees" was a company named Genetic Therapy. It was the first company to apply gene therapy to humans; that is, correcting genetic defects, particularly cancer, with the use of modified genes.

All of the biotech and medical devices in which Oxford has ever invested help solve some basic human problem. Most problems are quite serious. But there are some—well, they have more potential for humor than others. One that comes to mind is erectile dysfunction.

Early in the 1990's, we took advantage of the opportunity to invest in a small, non-public company that had invented a solution for impotence (this was pre-Viagra). It involved inserting a drug-soaked pellet into the urethra, a less-than-attractive delivery system. However, the results were worth it. Men in test studies indicated that the treatment would give them erections that lasted as long as four hours.

Needless to say, we believed we had a sure winner on our hands. The drug rocketed through FDA testing, which is normally an arduous process. I was quite amused to receive phone calls from senators and congressmen—I will not name them—on a regular basis, asking me, "You need any help with the FDA?" "Anything I can do to speed you through the FDA?" I was touched by the interest our elected officials took in the advancement of pharmaceutical science!

I remember the day when the drug completed phase III testing for the FDA (that is, formal testing on human patients). At Oxford, we sat anxiously by the phone, waiting for the call from the FDA. When it finally came, I was disappointed to hear that they had discovered a significant side effect of the drug. My heart sank. "What is it?" I asked.

"A sense of euphoria," came the reply.

As you might imagine, board meetings at this little company were very hard to endure with a straight face, due to the double entendres that were virtually impossible to avoid. We might start a meeting by asking for a report on any perceived competition for the drug, and the reply would be, "We expect the competition to be stiff." And everyone would dissolve in laughter.

Despite our high hopes, the company's prospects took a nosedive with the introduction of Viagra. Taking a pill is so much more attractive than delving into one's urethra, after all. But with the reported possible side effects of Viagra, the company's fortunes may be on the rise after all.

You see what I mean about unintended double entendres.

xxxxxx

Even with the best due diligence in the world, investing in any business is a roll of the dice. After all, ninety percent of all start-up companies fail within a ten-year period. Ten to twenty percent of companies financed by venture capital fail while venture capitalists are still investors.

Having said that, seeing a company die is something I've never gotten used to. Even after several years in venture capital, losing a portfolio company seems like a death in the family. Many venture capitalists will tell you that failure is always due to a mistake by a company's management. I don't think it's that simple. I've never seen two companies fail for exactly the same reason.

Ironically enough, both my greatest failure—and biggest success—in biotech venture capital thus far were closely intertwined. I never would have found the success without investing in the failure.

Here's what happened.

In 1988, my friend and future partner Jonathan Fleming told me that he thought that the science of genomics was going to become big business. Genomics is the science of hunting down genes; that is, identifying which precise genes in DNA are responsible for exactly what.

It's probably worthwhile taking a moment to explain what exactly DNA is. Every cell in the human body, other than red blood cells, has a nucleus that contains 23 sets of chromosomes. There are about 500 different cell types that have chromosomes. These chromosomes are made up of long strings of DNA. Each ribbon of DNA would apparently be six feet long if it were unwound, but they are so tiny—only one molecule across—that they are just detectable with an electron microscope.

The DNA ribbon itself consists of four "coding" units, which are given the abbreviations A, T, G, and C. On each chromosome, there are about one billion coding "bits," each of which can be either A or T or G or C. These coding sets are the basis for genes.

In all of the chromosomes, there are a total of about 150,000 genes. But these genes represent only about two percent of the information in the DNA. The remaining ninety-eight percent of the human genetic code appears to have no function—at least none that we know of yet. It's called "junk DNA."

The theory is that the "junk DNA" is a remnant of now-unused functions that represent the history of how we evolved from primitive organisms. That's why, for instance, about sixty percent of the genes found in yeast are also found in humans. And over ninety percent of the genes found in the brain of a fly are also present in human brains. (Think about *that* next time you've got the swatter out!) I am often amused to read about fundamentalists struggling with the concept of man evolving from apes (with whom we share ninety-eight percent of our genes). Imagine their frenzy if they focused on flies or yeast!

Now, having the genes for something or other doesn't in and of itself mean *anything*. That's because in any one of the 500 different cell types in the adult human body, about 30,000 of the genes are switched "on," and the remaining 120,000 are turned "off." When a disease occurs, genes that are normally switched off are switched on, and vice versa. This initiates the disease process. Similarly, when certain genes start to malfunction, aging occurs.

So it's pretty obvious that an understanding of the genetic process of how normal and abnormal tissue functions is the first, best way to figure out how diseases work and potentially how to prevent or treat them effectively. And that's what genomics is all about.

Now when Jonathan Fleming first talked to me about genomics in 1988, it wasn't on *anybody's* investment radar screen; it would be two years before the National Human Genome Program got going. It turns out that Jonathan knew a professor who claimed he could find genes quicker and better than anybody else. Because nobody else was on the genomics bandwagon, we were quickly able to form a company. The company was what I'll call "Gene Finder," and we put together three other venture capital groups to co-sponsor the company with us.

We set up Gene Finder near a very prestigious Ivy League university, to take advantage of the talent at the school. It paid off because we rapidly took on a dozen or so scientists and technicians from the university. The company also had a Science Advisory Committee—most biotech companies do—and its head was someone I'll call Professor Sam Skunk, from the southwestern United States. He had data on the Mormon population, which would ostensibly give us a method of identifying genetic propensities.

The reasons the Mormons were such good "guinea pigs" is that the Mormons are a discrete group—they tend to intermarry—and what's more, they tend to keep family medical histories, so that genetic traits can be traced for several generations. On top of that, their non-smoking, non-drinking habits eliminate certain non-genetic variables that prevail in non-Mormon populations.

This was all well and good, but growing Gene Finder was going to require convincing large pharmaceutical companies that they could make big money out of gene research. After all, they were going to be Gene Finder's customers, because they'd be the ones to exploit gene research commercially. And the big pharmaceutical companies just weren't biting yet.

We endured three or four frustrating years of demands for bigger pieces of the pie from Professor Skunk, increasing costs, and continued cash infusions from all the backers. We finally convinced one of the biggest pharmaceutical companies in the world that they should invest in Gene Finder because it would be useful to them if Gene Finder pinpointed the genes related to breast cancer.

Around the same time, Professor Skunk resigned from the advisory board. He was instantly recruited by a small-time venture capital player from Utah whom I'll call Hamden, and they together seduced the big pharmaceutical company into signing a deal with the new Skunk/Hamden company instead of with Gene Finder.

This was clearly a disaster for Gene Finder. There was nothing to do but pull the plug on the company, in which we'd invested a great deal of time and money—and confidence. At one point I had on my desk a thick document, a lawsuit against Skunk/Hamden for over $10 million. Clearly by leaving Gene Finder as he had, and undertaking the deal with the big pharmaceutical company, Skunk had opened himself up to a massive (and very winnable) lawsuit.

But I decided against. It would have taken two or three years of depositions, testimony, battling back and forth with lawyers to pursue the suit—time I wouldn't be able to spend raising money or investigating new companies. And once the lawyers had taken their sizable slice of the pie, there'd be very little to split between the four venture capital backers. Furthermore, that particular Oxford fund had basically closed; it was at the end of its ten-year life. It didn't seem to me a good idea to keep it open just for this lawsuit, or to siphon off management fees to fight the legal battle. So even though Gene Finder was my most significant battlefield casualty, I left the matter alone.

The Gene Finder story does have a happy ending, in the form of my greatest triumph thus far in the biotech venture capital business. From the ashes of Gene Finder arose the phoenix of Human Genome Sciences.

During the last few months of Gene Finder's life, I had been asked to join a debate team of patent attorneys to express the view of the venture capital community on the patentability of human genes. The person who called for my help was Reid Adler, Patent Counsel for the National Institutes of Health. He had filed the first patent for a set of human genes and gene fragments, and because of this he was under heavy fire from a lot of people in the federal government—the Waxman Committee, the Secretary of HEW, and many others. In the midst of this hue and cry, he called for help. "Get down here, Alan," he said, "and please say something sane about the situation!" So I beat it down to Washington to do what I could to help out.

All of the ruckus about patenting human genes had come about as the result of the work of an incredible person, Dr. Craig Venter of the National Institutes of Health. He had come up with a method for finding genes extremely quickly. And in doing so, he had created a national furor. Based on work generated by Craig's lab, the NIH had filed patent papers for thousands of partial gene sequences. That's what spawned the debate I'd been asked to join. This debate turned out to be a much bigger deal than I'd anticipated. The auditorium where we had the debate was stuffed to overflowing, and the other debaters included not just patent attorneys but also Bernadine Healy, the Director of the NIH.

After the debate, Craig introduced himself to me. From then on, we talked and schemed frequently.

The reason Craig's work was so compelling to me as a venture capitalist had to do with the way the genomics industry was shaping up. The United States Supreme Court had decided that genes were patentable; that is, if you were the first one to identify a gene, you could patent it, and anybody who wanted to use it for research—for instance, to find drugs that would treat the condition the gene controlled—would have to pay you for the right to pursue it.

This made the entire genomics industry a race to the finish line, and whoever came up with the fastest means of identifying genes was clearly going to win.

And Craig Venter looked like a sure winner to me. It was immediately obvious that his work was not only going to vault past Gene Finder, but that it was going to rock the world. Craig also recognized this, and as a result wouldn't join Gene Finder; he wanted his own deal, and a big one at that. He'd already been approached by Amgen, one of the biggest companies in biotech. They offered him $100 million to move his group to California. He didn't take the bait. He was in a situation with which I could identify. On the one hand, he wanted to stay in research—he wanted to win a Nobel Prize. But on the other hand, he wanted the financial fruits of his work.

What Craig really wanted to do was form a not-for-profit entity whose work product would be the property of a "virtual" side-by-side commercial company. The problem was the price tag—he wanted at least $60 million for this, way too much for any single venture capital firm to cough up.

Instead, I told him I would introduce him to the only group I could think of that was large enough to finance a large chunk of the project. Perhaps more importantly, it would have the vision to see how Craig's dream organization might work. The group was HealthCare Ventures, run by Wally Steinberg.

Wally Steinberg was a remarkable guy, either loved or hated by everyone who knew him. He was arrogant, egotistical, and had an overpowering personality, but at the same time he was a visionary. As Chairman of HealthCare Ventures, the dominant life sciences venture fund at the time, his word was essentially law. After he interviewed Craig, he astonished everyone by offering Craig $80 million for a non-profit center for Craig to run.

Wally's partners thought he was crazy. This represented about one-third of their assets—and they had virtually no chance of a return on their investment! But Wally was very clever. This financing would take the form of a commitment from a start-up, Human Genome Sciences, instead of HealthCare Ventures itself. This way if HGS failed, it was unlikely that Craig would get the $80 million.

But it didn't fail, and Steinberg's incredibly gutsy move paid off handsomely. Unfortunately, he didn't live long enough to see the great success of HGS. He died in 1996, only about eighteen months after it went public. Since then the company has grown very prosperous.

Back to Wally's $80 million offer to Craig Venter. Through Wally's master stroke, before long Craig Venter had a non-profit entity, The Institute for Genome Research (TIGR). Human Genome Sciences was the commercial operation, and Oxford was an investor in that. (I arranged for HGS to buy the remaining assets of Gene Finder—not that there was much left to buy after Skunk's flight.)

We started HGS with $10 million in venture capital money and the rights to any human genes sequenced by Craig's TIGR. Not even six months later, SmithKline Beecham, the giant pharmaceutical company, offered $100 million for 10% of the company!

Within a year of its startup, HGS was set to go public. It was an incredibly exciting day, and by coincidence happened to fall on the date of a preplanned meeting of the board of directors. The board members—of whom I was one—sat anxiously waiting for news of the IPO.

At 9 A.M., we got a call from Fred Frank, Senior V.P. of Lehman Brothers, one of the co-underwriters of the IPO. "The offering went very well," he said. "We expect it to go up from the twelve to fourteen dollar offering price, and then we'll try and stabilize the market."

'Stabilizing' is what investment bankers do to minimize major fluctuations in a stock price. They do this by buying and selling shares for their own account. And even though the system has plenty of room for fraud, for instance by hyping a stock, it generally works fairly well.

For instance, let's say a stock is selling at $10 a share and a seller wants to unload 100,000 shares, and this is such a huge transaction that the trade can't be matched up with a buyer who wants that many shares.

There are two choices: let the market sink until buyers see it as a bargain, or have the banker buy the shares (perhaps at a discount) until a new buyer is found. By buying the shares for the investment bank's own account, the stock is 'stabilized.'

After the call from Fred Frank, we went on with the board meeting, with one eye to the market quotes, waiting to see what would happen. By 10:30 A.M., there were no quotes for HGS, and it was obvious to us that something had gone wrong. The phone rang. It was a Senior V.P. from Smith Barney, another of the underwriters. He told us, "We have a little problem." Uh-oh.

"Everybody wants to buy," he said. "Nobody wants to sell." Eureka!

Half an hour later, the stock started to trade at $16, and by the time the board meeting was finished the shares had shot up to $28.

On the third anniversary of the initial $10 million investment, HGS had a market cap of just under $1 billion.

It was the most successful biotech deal of all time.

Chapter Twenty-Eight

Climbing Mount Kilimanjaro

The idea of a mid-life crisis is something I never took seriously. The middle years of my life seemed to me to be one long crisis of some kind or another.

Nonetheless, turning fifty in 1986 was a watershed moment for me. I resolved to do something I had long put off: Get a complete physical.

It took me almost two years to actually go through with the physical. Frankly I didn't want to know the results, largely because I took a fatalistic approach to health. Of course I knew as well as anybody else about all the reams and reams written on the subject of eating right and getting exercise. But as a trained scientist I had always considered nutrition at best an inexact science. I figured that most diet books were written by cranks and kooks.

As for exercise, my attitude was that it was all right for three kinds of people: those who had the time (I didn't); masochists (I wasn't); and those who didn't appreciate comfort (I did). Furthermore, I knew several men who'd died of heart attacks while exercising, guys who were ostensibly very fit. They'd have been better off on the couch with a remote control in one hand and a bag of Doritos in the other, I reasoned.

In any case, I was convinced that health was determined far more by genetics than by diet and exercise. My mother at the time was in her late eighties (she lived to be 92). My father had died accidentally at age 76.

So even if the results of my physical were bad, I didn't believe there was much I could do about them. Why bother knowing at all? And on top of that I didn't want to be lectured by my doctor. I didn't need to hear how unhealthy my haphazard meal plan and penchant for snack foods were. I couldn't bear the thought of giving up my traditional Sunday

English breakfasts: fried bacon, fried sausages, followed by eggs, tomatoes and mushrooms fried in the same bacon fat, and topped off with fried bread. I didn't want to be told that years of a high-stress, low-exercise, unhealthy-diet existence had taken a toll on me.

But of course they had. My blood tests showed that I had too much of the wrong kinds of cholesterol, and not enough of the right kinds. My blood pressure was increasing significantly. I was twenty-five pounds overweight. The exercise regimen I considered appropriate—a game of tennis every week—clearly wasn't enough.

I could have ignored the information. But I couldn't, not really, in the face of my deteriorating health. It was time to do *something* about it.

I wasn't interested in anything too drastic. I decided that I could probably tolerate replacing sugar with NutraSweet, and I could live without chocolate, cheese, meat, and high animal fat products. Not *happily,* mind you, but I could do it.

To my astonishment, the results were dramatic and immediate. Within three months I had lost thirty pounds, and was down to my college weight—158 pounds. While my "good cholesterol" was still abysmally low, my total cholesterol and "bad cholesterol" had dropped by half.

My doctor's verdict was guarded. "Well, it's a good start," he said. "But the only way to continue your improvement is through"—the dreaded word—"exercise." Ugh.

My response was to double my tennis playing to two games a week. I also bought an electric treadmill. I used it on those occasions when I could find no excuse to avoid it.

What was beginning to dawn on me was what chronic dieters have always known: it is much harder to maintain a diet and exercise program than it is to start one. After the first flush of success on my diet, my interest began to lag. And I pinpointed exactly why. There's no goal in dieting. It's not as though you're working toward a specific end. *Maintenance* is uninspiring. I needed to have a brass ring to grab for.

And before long, my friend and tennis partner Bill Kokot provided one.

In late summer 1988, Bill raved about a hot-air balloon safari he had taken, gliding over the Serengeti Plain in Kenya. While he was there, he'd met a party of hikers who had just climbed Mount Kilimanjaro, the tallest peak in Africa. These climbers regaled Bill with stories about their adventure. He was instantly intrigued by the prospect of making the climb himself. "I looked into it, Alan," he said, "and travel agents here can arrange a tour. We don't need any mountaineering skills. It's basically just a very strenuous hike." *Very* strenuous indeed—Kilimanjaro is 20,000 feet high!

Now, *here* was an adventure that could give my exercise program just the jolt it needed. But even more importantly than that, what a thrill! It conjured up all of the romantic visions of Hemingway and exotic Africa. I remembered seeing Gregory Peck in *The Snows of Kilimanjaro* many years previously. And what better way to beat the ennui of middle age than with the excitement of this kind of a challenge?

The more we talked about it, the more Bill and I became enamoured with the idea of hauling our aging bodies up the highest mountain in Africa. It was very easy, at that point, to ignore the dark underbelly of the climb—altitude sickness, blinding headaches, pulmonary edema, extremes of heat and cold, total physical exhaustion and what were certain to be very primitive conditions. We were swept up in our fantastic visions of what life lived to the limits could be.

My interest in going was spurred all the more by a book that very strongly influenced me at the time. It was called *Big Is Invisible,* and it was written by Belinda Charlton, the stepmother of my friend Randal Charlton. The book was the story of her journey from being a 300-pound cancer victim to a svelte, healthy woman who competed in the London marathon and became a TV celebrity in the process.

While the challenges she faced were monstrous compared to my rather paltry health problems, I somehow made the correlation between Belinda's efforts to overcome her handicaps and my own much less challenging effort to achieve health and happiness by climbing Kilimanjaro.

By January of 1989, Bill Kokot and I had decided we were definitely going to go for it. What sealed our decision was the experience of a local chap. He was fat, a smoker and a drinker, and he claimed he had made it almost to the top of Kilimanjaro, and he went on to add that 60% of the people who try it actually make it almost to the top. "Whenever I felt tired," he said, "I sat down and had a swig of whiskey from my flask." Well, if he could make it, Bill and I concluded, we could get there hopping on one leg. (It wasn't until much later that I found out that this guy worked for Lindblad Tours—the people who arrange the very expensive excursions to Kilimanjaro. On top of that, his 60% figure was somewhat fanciful; the author Michael Crichton claims the percentage who summit is closer to 14%. Having made the climb myself, I agree with Crichton.)

While I had been discussing the possible trip with Bill, I'd been chatting about it with other friends as well, including a fellow Brit, Michael

Irving. Michael was whippet thin, in excellent trim, and as he was a few years younger than Bill and I, he figured he could make the climb with no problem.

The three of us agreed that we'd fly over in September of that year. That allowed for seven months to train.

Now, *here* was a goal I could sink my teeth into!

It wasn't until we'd made a nonrefundable down payment on the trip that the enormity of the venture started to envelop me. The first cold slap of reality came when I tried to ramp up my exercise program. I knew I'd need to start using my treadmill some more. So I resolved to run about a mile each morning, before going to work.

Now a mile isn't much. But this was the worst part of my whole training plan. I am not an early morning person. And on top of that, the treadmill was in my cold, dank basement. So while the rest of the household was asleep, I had to motivate myself to descend into the dungeon and run in place, like a rat on a wheel.

But I kept up with it. I quickly got frustrated because no matter how hard I worked, I couldn't seem to beat a seven-and-a-half minute mile. I also started to realize that there wasn't likely to be much correlation between putt-putting along for brief stints on a flat treadmill at sea level, and climbing for many hours at high altitude.

Bill and I deduced that purchasing a mechanical climbing machine would be better for strengthening the leg muscles we'd need for the climb. We figured this kind of climbing would be a cinch. After all, we both had strong legs from our frequent tennis matches.

Hah! No sooner had my StairMaster been installed, than I discovered that it put excessive stress on my knees. I got rid of the thing pretty quickly—as did Bill—but the knee pain continued until shortly before we left for Africa.

As the spring of 1989 slipped away, we occasionally came across magazine and newspaper articles about the rigours of climbing "the big K." There were four stages of the climb altogether. You started at 6,000 feet, and went to 9,000 feet for stage 1. The second stage took you to 12,000 feet. The third, to 15,000, and the fourth, to the summit. Each stage would take between four and seven hours.

These figures threw me into a depression. For a start, they made my daily ten minutes' heavy exercise look laughable. For another, I had never walked or climbed *anything* for four hours in my adult life. I'd *avoided* it. And even if I hadn't, it seemed to me that four hours at low altitudes would bear no resemblance to the high altitudes we'd face on the mountain.

I immediately changed my training philosophy.

Because I live in a hilly part of Connecticut, I decided that getting outside and power-walking or jogging uphill as far as I could would do me more good than hanging out on the treadmill in the basement. Now mind you I didn't relish this. I have never regarded walking as a worthwhile pastime. I thoroughly agree with the saying, "If God had meant us to walk, he would never have let us develop automobiles." The pleasure derived from observing nature has always, for me, been outmatched by the discomfort of sweating and blisters.

Nonetheless, with September looming closer and closer, it was worth a try. Initially E.J. drove me about a mile from home, and I would walk or jog back to the house. That mile might as well have been a thousand miles at the beginning. I made it through by walking up the hills, jogging back down, and counting my steps to take my mind off my discomfort.

Unlike the treadmill—which seemed never to get easier, no matter that I was still faithfully plodding away on it every morning—my outdoor hikes rapidly became less tedious. I could move more quickly and cover longer distances.

Eventually, about a week before leaving for Africa, I found that I could jog or power walk up to twelve miles, at an average speed exceeding five miles an hour. More importantly I could do it without any overwhelming discomfort. Even with climbing boots and a backpack, I could cover the same distance at about three and a half miles an hour. My confidence increased with every step.

These long treks had an interesting effect on my state of mind. For the first few days, I was acutely aware of my surroundings. Every footfall was obvious. But I quickly moved to a dream-like state, lost in thought for most of the walk.

Whether this was the "runner's high," the result of endorphins kicking in, or a kind of self-hypnosis induced by my rhythmic steps and counting, I couldn't say. But it became clear that I was in some kind of trance one day when I was stopped by a motorist asking directions. I knew exactly where he wanted to go. But for the life of me, I couldn't spit out how to get there. I stood staring at him, in a fog not only as to which way he ought to go but unable to identify exactly where I was standing. I was like a mouse in a maze, mechanically following a rote path home. The guy shook his head and drove away.

I was to be even more embarrassed on the last day of my training. E.J. dropped me about five miles from home, at a spot where she'd let me out many, many times before. I almost immediately clicked into zombie brain. But this time, I took a wrong turn and got lost. I eventually covered about seven or eight miles before I gave up and hitchhiked home. For all I know, the guy who gave me a ride might have been the same one I had previously misdirected.

Despite my jogging vapor locks, I convinced myself that I was fit enough for the trek. It was time to focus on what I ought to pack, in terms of clothes, food, and anti-altitude sickness drugs.

As I scanned the guidebooks for advice on what to take, I learned more about the history of Kilimanjaro. Prior to World War I, it was in the British East African Protectorate (later to become Kenya), close to the border with German East Africa. At the end of the 19th century, in a move of sheer imperialist *droit de seigneur*, Queen Victoria "gave" Kilimanjaro to the German Kaiser as a birthday present. Ironically, the British soon got the mountain back. After World War I, the lands of German East Africa were confiscated by the British. Kilimanjaro was then part of the Protectorate of Tanganyika. After its independence in the 1950s and 60's, it joined up with the island nation of Zanzibar to form the republic of Tanzania.

The mountain, the highest in Africa, consists of a flattened volcanic dome. The two main peaks on the tourist route are Gilmans' Point at 18,650 feet, and Uhuru Peak, at 19,300 feet. Although the mountain is virtually at the equator, it is very cold. The peak is snow covered year-round. And the air is *very* thin. By comparison, pilots generally require oxygen to function above 10,000 feet. The top half of the climb would be above that altitude.

On top of the possibility of altitude sickness, we would face the risks of all kinds of nasty illnesses. It made absolute sense to be fortified with vaccines in the U.S. before we left. On a prior trip to Africa, when I visited Kenya in 1977, it wasn't until I got to the Nairobi Airport that I realized I'd forgotten to get a cholera shot in the States. Nobody checked, and I got through, unvaccinated and disease-free.

I later realized how lucky I was. I heard stories that many diseases were prevalent in Kenya and Tanzania, and even if the diseases were dormant, the border authorities would still give you an injection—evidently changing needles once every few hundred shots!

Bill, Michael and I had no intention of going through *that*. So at appropriate intervals we all took yellow fever, cholera, tetanus and typhoid shots, as well as pills for malaria type I and II. Bill and I also elected to get gamma globulin shots to increase our immune resistance.

Michael didn't. The shots were reputed to be only 60% effective and were derived from human blood, thus carrying the possibility of AIDS infection. Michael felt the reward didn't outweigh the risk.

As it happens, the week after I got a gamma globulin shot, I felt distinctly unwell. I wondered whether Michael might be right after all. It turns out I had a low-grade infection, and was soon feeling fine.

Exotic diseases aside, what scared me most about the climb was the anticipation of altitude sickness. I had for a decade or so suffered from periodic, blinding headaches. They were debilitating when I was in the comfort of my own surroundings. The thought of trying to trudge up a mountain with cannons exploding in my head was virtually unthinkable. So I had to do *something* to avoid getting altitude sickness.

The best way to beat altitude problems is to spend a while acclimatizing to the altitude at intermittent stops. But we didn't have that luxury, and so we had to rely on other methods. While Michael decided to tough it out, Bill and I laid in a supply of Diamox pills. Although Diamox would supposedly do the trick, it was reputed to have a nasty side effect: it made everything taste nauseatingly sweet. In Connecticut, that seemed like a risk worth taking. On the mountain—well, that was another story.

xxxxxx

We spent our first evening in Africa at the Safari Club in Nairobi. At one time the club had been the bastion of British civility. But like much in Nairobi, by 1988 the club was deteriorating, a mere whisper of its earlier, haughty grandeur.

From Nairobi we were due to fly to Kilimanjaro Park via Air Tanzania. We were pleasantly surprised when the Air Tanzania plane landed at the Nairobi Airport. It was a freshly painted Boeing 727 with a sporty giraffe on the tail. Unfortunately, the inside of the plane didn't live up to its spiffy exterior. We later learned that this was Air Tanzania's sole jet aircraft, and it flew a continuous route between Dar-es-Salaam (the capital), Nairobi, Kilimanjaro, and back again.

Tanzania was every bit as primitive as we expected it to be. Most of the natives scratched out a hard-scrabble existence as tenant farmers. Others acted as guides for tourists, who visited either for safaris or for the climb up Kilimanjaro. Virtually no tourists ventured to the coast to visit Dar-es-Salaam. We heard reports that it was dirty, poverty-stricken, and crime-ridden, with none of the alluring plumage of a tourist destination.

Our first exposure to Tanzanians was at the Mt. Kilimanjaro Park Airport. The experience didn't show them off to their best advantage, but it wasn't entirely their fault. As is true in many countries I've visited, the local currency wasn't "convertible"; that is, if you had Tanzanian shilingi, you couldn't convert it into "hard" currency like dollars, marks, francs, or pounds.

As a result there was a tremendous black market in foreign currency, and it's all that the guides and porters would accept. It made them seem surly and rude. We'd known about the hard currency attraction before we left the States; the only problem was that you weren't technically allowed to bring in any "hard" currencies.

I managed to get a five pound note through customs by hiding it in my sock. Bill was almost arrested when the customs officials found he was carrying a couple of undeclared dollars. We only got by the airport customs guards by buying them off with a few dollars—a process that we had to use frequently. Indeed, we only got our baggage back out of Africa at the end of our trip by paying off the airline porters at the Nairobi Airport.

The ride from the airport to our hotel at the foot of Kilimanjaro took about an hour and a half, with the van bouncing along what could only in the most generous sense be considered roads. There were frequent craters and many sections were unpaved.

We didn't expect much of the Hotel Kibo, and as a result we weren't disappointed. It was essentially a way-station for anxious trekkers on their way up the mountain and the exhausted (and often dejected) trekkers on their way back down. We were to spend two days there, acclimating to the altitude—although at only 6,000 feet I'm not sure how necessary that really was.

The stay did give us a chance to meet other adventurers. We were very eager to hear of returning climbers' experiences. That is, at least, until we actually heard what they had to say. The first returning trekker we met was a twenty-three-year-old Australian woman named Elizabeth. We actually met her underwear before we met *her*.

Michael, Bill and I were sharing a room with a small terrace, surrounded by three-foot-tall brick walls. The morning after our arrival, much to our amusement we found female undergarments draped on one of the walls of our terrace, presumably left there to dry by our neighbor. As we were chuckling over this, Elizabeth came out to retrieve her lingerie. She wasn't at all embarrassed by our presence, and we had a pleasant chat.

It turns out she had been brought off the mountain the previous evening, having suffered from severely swollen knees. She was with a group who were intent on making it to the top. She told us they left her at the Ice Cave at sixteen and a half thousand feet and continued to the top. They picked her up on their way back down.

As she went on talking, our minds drifted. Although none of us voiced the concern, we were all thinking the same thing: If this fit young hiker couldn't make it all the way up, how the heck are three 50+-year-old guys going to do it?

The next person we met alarmed us even more. Our neighbor on the other side was a man named Al. He was in his early fifties—good news there—but he routinely ran the Honolulu marathon, and had worked out for years, running up and down Hawaiian canyons for training. He hadn't made it to the summit of Kilimanjaro. He was brought down after passing out from hypothermia.

It was quickly dawning on us that summiting had less to do with fitness than with the ability to fight off the elements, and most particularly altitude sickness. While Michael was going to tough it out without any drugs, Bill and I were quietly pleased about our secret weapon—the Diamox anti-altitude sickness pills. We started popping Diamox on Day One, at the foot of the mountain. It immediately gave me a tremendous sense of confidence. But there was no question that its side effect was overwhelming. It made everything taste so sweet as to be inedible, and that included even water and seltzer.

From the first day of the trip, I was reduced to eating virtually nothing. I consoled myself with the thought that the meat we were provided was probably rat burgers, and so I wasn't missing much. The only thing I could choke down was trail-mix, a mixture of dried fruit and nuts. But I hadn't brought much of it with me from the States, and I was soon to regret that.

xxxxxx

Sunday, September 17, 1989. We were due to meet our guides and porters at 8:30 A.M. We woke up early, excited beyond measure, and quickly ate breakfast and packed everything we needed for the climb into tarpaulin bags. We left our valuables in the hotel safe and our remaining luggage in a "safe" room at the hotel. We considered having "one for the road," but the altitude medicine made everything taste so vile that we decided against. The only liquid sustenance Bill and I could stomach was lemonade. So we took lemonade powder in the one-quart water bottles we were carrying. At that early stage of the trip, we were still adding purifying tablets to the water, to chase away any nasty bugs.

The climb was supposed to take three and a half days up and two and a half back down, for a total of six days.

While Bill and Michael donned their climbing boots from the beginning, I decided to climb for the first two days in tennis shoes, since I didn't trust my boots not to give me blisters. My plan was to wear a tennis shirt and light sweater plus jeans, the type of wear that I had trained in back home. I knew that I would sweat a lot, and I had planned three sets of gear for the climb. I would change for the first two days, and then wear the remainder for three days, peeling off layers as we came down the mountain. Of all the plans I made, this one actually worked quite well.

At 8:30, our van arrived. We met John, our Lindblad rep, a large black guy, very fluent in English, very competent and a skilled driver. He immediately instilled confidence in us. With him were Bariki, our chief guide, Dennis, the assistant guide, and a motley crew of five porters.

Bariki was black as tar, bearded, and wiry. He could only have weighed about a hundred pounds after a full meal. I thought that anyone who climbed mountains frequently would have to have a solid layer of fat, but Bariki proved that wasn't true.

He looked to be about our age, but as the veteran of many treks the stress of climbing might have aged him. Dennis, on the other hand, seemed to be thirtyish, well-built and strong. He spoke English much better than Bariki, who had a working vocabulary of about a hundred words.

The eleven of us squeezed into the nine-seater van, with all of our baggage crammed in as well. We set off for the half-hour drive to the Kilimanjaro Park gates. On the way, we stopped at the Kilimanjaro equivalent of a Seven Eleven to pick up food supplies for the trip. The three of us, despite peering out the windows, didn't get a good look at what we were to eat. That was probably a good thing. At the store, we also picked up another passenger—a rather substantial young lady who appeared to be the girlfriend of one of our entourage. The badly overloaded van groaned to the Park entrance, where we disembarked.

We had anticipated a long, drawn-out registration process. Instead, we zipped through in no more than twenty minutes, because we'd arrived early enough in the day to avoid long lines.

The Park entrance was a fortified main gate guarded by armed police. While the police may have been there to stop potential bandits from creeping in, I think their presence had more to do with making sure everybody paid the $180 entrance fee. They were also responsible for doing a body count, ensuring that for every person who went in, someone came out—dead or alive.

We were held up a few minutes when the porters, in unloading our bags, discovered that a box of eggs had broken. It must have been a big deal, because they went back to the store to buy a replacement. The thought of eggs as a staple of our diet while I was on the nausea-inducing Diamox pills was not pretty, but I dismissed it at the time.

At the trailhead there was a large sign, listing the various stages of the climb. Apparently this was a traditional photo opportunity, and so we stopped for an obligatory picture. Michael went first, brimming with confidence, and had his picture taken pointing to Uhuru Point on the map, at the summit. Bill followed suit, pointing to Uhuru. When it was my turn, I pointed to Mandara—the lowest point on the climb. Everybody thought this was very funny, but I did it only partially in jest. My apprehension was rising.

We took a few minutes to study the sign, and that buoyed my spirits somewhat. The times that it suggested were necessary for each leg of the climb seemed eminently manageable. Three hours to Mandara, which was at 9,000 feet. Five hours from Mandara to Horombo, at 12,000 feet. Another five hours from Horombo to Kibo, at 15,000 feet. Yet another five hours to Gilmans' Point, at a lofty 18,650 feet, and an hour and a half beyond that to Uhuru, at 19,000 feet. Surely this was all achievable! But my mind harkened back to Al, our Hawaiian friend. He said it had taken him eight hours to get from Kibo to the Gilmans' peak, three more than the sign stated. How accurate *was* this sign, anyway?

We started our trek. Al had warned us that the first day on the mountain was fairly tough going. Forewarned, we were hardly fazed to find that it was, in fact, a strenuous climb. This first leg took us through what was billed as a "rain forest." It was certainly lush, although virtually devoid of animal and insect life. We had been told that if you get there early enough, you disturb hordes of monkeys. We found nary a chimp, only the occasional exotic bird. And despite the "rain forest" moniker, it didn't rain, something we had perhaps guaranteed by carrying our rain gear in our backpacks.

We arrived at the Mandara checkpoint in excellent spirits. Although we'd sweated quite a bit, the temperature had been more than comfortable, in the low seventies. Best of all, we'd made it in 3¾ hours. That was only 25% longer than the posted three hours. I quickly calculated that if we maintained that rate, the climb from Kibo to Gilmans' Point should take us only 6½ hours, far ahead of Hawaiian Al's eight-hour pace! We were quite full of ourselves, and believed the day augured well for things to come.

I should have stopped doing math at that point. But I didn't. It dawned on me that I had gotten used to judging distance traveled instead of time and energy exerted. In my "power" walks at home, I was able to do about 4½ miles an hour on the flats and 3½ going uphill. Standing at Mandara, I realized that because we'd been walking slowly to adapt to altitude, we were averaging about 2 to 2½ miles an hour. That meant that we'd covered about eight miles, compared to the 4½ miles listed in our literature!

Figuring that the climb to Gilmans' Point was supposed to take 17 hours, adding 25% to reflect our actual time, and multiplying by our two-mile-an-hour rate, it was shaping up to be a forty-mile hike up the mountain. This confirmed some of the literature I'd read, which said that the total actual distance up and down the mountain was about seventy-six miles. Since we were supposed to make the trip in five days, we'd have to average fifteen miles a day. Under strenuous conditions. At high altitude. I was dismayed as I compared this to my training schedule, which never exceeded twelve miles a day, at sea level, under mild conditions at best.

I was starting to get worried. But I didn't share my thoughts with the others. And our actual stay at Mandara wasn't too bad. The stopping point consisted of a series of A-frame huts, measuring about ten by seven feet at the base. They were each meant to sleep four people, but Michael, Bill and I got a cabin to ourselves, largely I think due to Bill's ingenuity. As a building contractor he was our master negotiator, and in this particular situation I would imagine a little graft was involved. In any case, we got our own cabin, and the fourth bunk held our baggage.

Bariki and his crew prepared our dinner and served it to us. It looked pretty good, soup and a side dish, and Michael tucked in greedily. Thanks to the Diamox, Bill and I couldn't eat a bite. In fact, the sweet, sickening smell of the firewood used by the porters to cook everything was to grow in nauseating splendor as we headed to the higher altitudes.

The hike to Mandara had taken us less than half a day and we weren't due to leave the stopping point until the following morning, so we were left with quite a bit of time on our hands. We swapped stories, talking about our lives, ambitions, philosophies. The setting seemed to inspire this kind of soul-baring. Had I been another kind of person, I would have said we spent the time "bonding." But I'm not that kind of person.

The following day dawned fine and only slightly chilly, even though at 9,000 feet we were near the cloud line. Since it had not rained in the rain forest the previous day and there was not a cloud to be found in the sky, the guides assured us that we wouldn't need to carry our rain gear. The porters packed it into our bags and hauled it all up ahead of us. The plan was to head for Horombo, which at 12,000 feet was supposed to be a five-hour hike.

Needless to say, the fact that we didn't have our rain gear meant that soon after we started out, the sky darkened and it poured and poured and poured. We were soaked in short order. Although the terrain was described as "moorland" in our literature, we were too miserable to appreciate it. We numbly squished up the trail, plodding behind each other like pack mules.

It was late afternoon by the time we reached Horombo. It had taken us some seven hours to get there from Mandara, compared to the five it was supposed to take. We were exhausted and more than a little depressed. The water had served to effectively dampen our spirits. On top of that, it was cold, so cold that we had to pry Bill's frozen fingers from his walking stick.

The huts at Horombo were much like those at Mandara. There was a structure designed as a main eating area, although Bill and I didn't feel much like using it, the Diamox working its nausea-inducing magic once again. Much of our gear had arrived damp. With our exhaustion, lack of food and the permeating rain, it looked like the third day was going to shape up badly. Had I not taken a sleeping pill, I doubt that I would have slept.

Sure enough, I woke up in a foul mood, but felt better as the day wore on. We were headed to Kibo, which at 15,000 feet was supposed to be another five-hour hike's distance away. The terrain was described as "alpine desert" and we were above the clouds, but we weren't going to tempt fate again. We wore our rain gear, so of course it didn't rain a drop.

At 13,500 feet we passed the vegetation level at Mwenzi Ridge, and the last water point shortly after that. From then on we were lugging our own water, and it must have been very heavy for the porters, since we were each drinking four or five quarts a day. Whatever pleasure I had gotten from the hiking had thinned with the altitude. With each plodding step, I counted to myself, "One . . . two . . . three . . . four . . . " I have never counted to one hundred so many times in my life.

Up to this point, Michael had clearly been the best off of the three of us. He was the youngest and healthiest. He'd eaten plentifully at each stop, and had slept peacefully. Bill and I had stuck with the Diamox, which had meant that we'd eaten scarcely a bite in two and a half days. But at 14,500 feet, Michael started to feel ill. His face turned gray. He discreetly threw up a few times. He was clearly being overtaken by altitude sickness, but he didn't want to quit.

Bill and I between us managed to carry Michael's knapsack the last 500 feet to Kibo, where we arrived in late afternoon. Once again, an ostensible five-hour hike had taken us almost seven hours.

What was more alarming was that as we approached Kibo, we'd begun to notice gravestones. I pointed this out to our guide Dennis, and he said, "Oh, those—those are for porters who died on the trek." I was suspicious of this when I noticed some stones marked with signs of German and Japanese tourists.

Apparently the single existing Tanzanian helicopter could not get above 12,000 feet for rescues. That meant that if you dropped dead on the mountain, you stayed on the mountain—dead.

So we trekked toward Kibo, Bill and I hauling Michael's knapsack, Michael gray and ailing, passing the gravestones of previous trekkers. By the time we reached Kibo, we were having a hard time remembering each other's names. The lack of oxygen was causing a common side effect: we were becoming disoriented. I had read that people often fall off cliffs at high altitude because they forget where they are and where they're stepping.

We began to wonder if we were going to make it out alive.

The last part of the climb from the Kibo Hut was scheduled to begin sometime after midnight. Because we arrived late, this would mean that we would only get four or five hours of rest before we had to set out again. On top of that, our late arrival meant that the choice of bunks was limited.

Unlike the individual hut setup lower down on the mountain, at Kibo Hut there was basically one big room with a couple of dozen bunk beds. There were already twenty people in the hut when we arrived. Michael and Bill headed for upper bunks at the far end of the long narrow room, reasoning that it would be quieter and warmer, being far from the opening and closing door. I was left with a lower bunk right next to the door.

Sure enough, people were constantly coming and going. I felt a little sorry for myself, but it turns out my bunk was a lucky choice. The hut had little or no ventilation—not to mention oxygen—making life in the upper bunks intolerable.

Shortly after midnight, Bariki and Dennis roused us. They gave us some food—a waste of effort for Bill and me—and we got underway by one A.M., Bariki and Dennis leading the way.

We were soon in trouble. Our condition hadn't changed much from the time we arrived at Kibo. Michael was very sick. He'd been throwing up all night, but because he hadn't complained at all we underestimated how serious his condition was.

I was flat out spent. I tried a variety of means of trudging, trying to preserve what little energy I had left. I settled on a system of breathing in for three steps, and breathing out for the fourth. I stopped for a rest every thirty or forty steps, catching my breath—what there was of it—for a minute before starting out again. One, two, three, in—pfssshhhh, out.

Although we had planned to approach the summit under a full moon, we'd missed it by one night. Nonetheless the moon still loomed large, and the sky was clear and beautiful, with moonlight illuminating the pathway before us. But we couldn't appreciate it. The air was bitingly cold. We were each wearing everything we had left that wasn't wet. I wore a thermal undershirt, shirt, high-neck thin sweater, heavy sweater, and ski parka. Underpants, long johns, jeans and ski pants. Three pairs of socks and thermal boots. Balaclava helmet, ski hat and goggles. Gloves covered with an extra pair of socks on my hands.

My water bottle, full of a quart of water and the last of my lemonade powder, was strapped to my belt, under my parka.

It froze.

Climbing was a delicate balance between moving, exhaustion, gasping for air, and stopping to regain some energy while freezing. I can only imagine how the others felt; we didn't discuss it. Michael shouldn't have started up from the Kibo Hut at all. After half an hour, he clearly could go no further. Dennis took him down the hill and back to the hut so he could rest and hopefully regain his strength.

Bill and I stuck it out for another hour with Bariki, making painfully slow progress up the mountain. Bill sounded increasingly awful. His gasping for breath reminded us both of his heart fibrillation two weeks before we'd left for Tanzania, brought on by the shock of cold liquid on his dehydrated system while he was doing heavy work on his son-in-law's driveway.

I was torn. On the one hand, I did not want to be left alone on the mountain, which would have happened if Bariki had to take Bill back down to the Kibo Hut. But on the other hand I had visions of Bill suffering a heart attack, and having to carry his body back down the mountain.

As I was lost in thought, a guide came down the mountain toward us, leading a Japanese trekker who'd been overcome by mountain sickness. Much to my relief Bill decided to return to Kibo Hut with them.

Before he left, Bill told me, "Michael and I will wait until daylight tomorrow for you, before we go back down the mountain." At the time the comment didn't register properly with me, my disoriented brain unable to process much of anything except breathing, plodding, breathing, plodding.

So it came down to Bariki and me. Things improved briefly. Without the others to worry about, I could set my own pace. For a time, I convinced myself that I would go on for another few minutes and then return to my friends. This kept me going for a while, since the "few minutes" were never up. Unlike the reports of Michael Crichton and others that I'd read about, it wasn't the cold that got to me. The problem was that my muscles were completely unresponsive, and I was almost totally sapped of energy.

About half an hour after Bill turned back, Bariki pointed out a light in the distance, and up at an angle of about thirty degrees. "Ici Cava," he said. "Ici Cava." This was the infamous Ice Cave! At that moment it seemed an incredible distance away, but it gave me a target to aim for.

The next hour and a quarter were about the worst of my life. Plod, plod, plod. I felt a growing sense of isolation and exhaustion. My morale had plummeted. But then—the cave! This meant we were at 17,000 feet. I slumped down inside.

"Ice cave" is certainly a misnomer; it sounds like one of those Scandinavian hotels that they carve out of the ice every year. Instead, it was hardly cavernous; it was basically an igloo, about seven feet in diameter. Bariki squatted down nearby, and I persuaded him to take a photo of me—a photograph I was not entirely sure I'd ever live to see.

I tried to warm up my frozen lemonade for a drink. I wound up crushing it as best I could, and sucking on the popsicle niblets. I eventually found enough energy to look around and take a few pictures, and only stopped when my lens froze. I peered out at the path that lay ahead to the summit. The terrain was gravelly rock, and the snow and ice line was still above me.

For a few minutes I was tempted to turn back. After all, what did I need to prove? I was beginning to feel distinctly unwell. I was dizzy and disoriented, and my body felt like a lead weight from exhaustion. With what little logic I could muster, I calculated whether it was worth going on. Hawaiian Al had taken eight hours from the Kibo Hut to the summit. I had climbed halfway in just over three hours. If he could do it, I could, too!

After a fifteen minute rest at the ice cave, we moved on. Not far above the cave came something I'd been expecting: scree. The best way to describe the terrain known as scree is to say that climbing on it is like trying to scramble up a pile of gravel. As you move one of your feet forward, the underlying gravel slides it back down almost to where it started.

Suddenly I got angry with myself. I hadn't taken the scree into account in my calculations. It was increasingly obvious that halfway to the summit in three and a quarter hours didn't translate to six and a half hours to the top.

My pace was slowing. For every thirty steps I took, I only made it ten paces forward.

I began to feel overwhelmed. My foggy brain figured that even if it took me six and a half hours to get to the top and another hour and a half to get back down to Kibo, it would be 9 A.M. before I got back, and my friends would have already left. I would be left to an eight-hour descent by myself.

I had now been climbing for five hours since Kibo, and I could only take five or six paces before stopping to rest. I had chest pains. I was utterly demoralized. At close to 18,000 feet, I sank to the ground in bitter frustration.

Bariki seemed to have an endless supply of energy. "Are you OK?" he asked. I simply could not reply. He shone his flashlight in my eyes, and must have seen the tears of frustration burning down my cheeks.

The darkness began to lift, and I could make out the narrowing of the mountain to the peak ahead. Bariki nodded toward it. "You must get up, or you freeze."

I momentarily panicked. I was not at all sure I *could* get to my feet. Never mind whether I could make it up or down the mountain!

Kilimanjaro had become my own personal monster. I *had* to conquer it. But in the sliver left of my rational mind I knew I had gone well beyond my physical limits. I thought I might die on the mountain. I began to think of my wife E.J. and my family and friends at my funeral, shaking their heads and muttering, "What a waste."

From the safe distance of hindsight, I now look back and romantically believe I could have made those last six or seven hundred feet. If only I could have rested after the climb to Kibo Hut. If only my friends were not leaving. If only I'd had a decent meal in the past three days. Heck, if only I'd had a few candy bars with me . . .

But realistically, in the shape I was in, exhausted, starving, disoriented, it wasn't possible.

In that moment, as I sat pondering my own death, I began to think dying wouldn't be so bad after all. It would be as comforting as sleep. It would save me from the pain of dragging one foot after the other.

Bariki must have seen all this before. He knelt down, squeezed my hand, and said quietly, "We go down now."

xxxxxx

"As easy as falling down a mountain" is a misnomer. It wasn't all that simple coming back down Kilimanjaro.

Bariki and I got back to Kibo Hut without incident. Bill and Michael were waiting, and pretty much ready to go by 8:30 A.M. As we headed down from the lofty altitudes, Michael started to feel much better.

At about 14,500 feet, Bill pointed back up the mountain, and said, "Hey, Alan, take a look at *that.*" As I swiveled my head to see what he was pointing at, I stumbled over a rock and tumbled a few feet down the mountain. As I fell I thought, 'How ironic—I made it *up* the mountain, but the trip back *down* might kill me!'

When I came to a stop, my knee was badly sprained. Fortunately a German physician was climbing the other way and was able to strap up my knee well enough that I could go on. Michael half-carried me and my equipment the rest of the way.

Around 12,000 feet, Bill suddenly collapsed. My first thought was: heart attack! But fortunately he had simply been overcome with exhaustion, and he revived after a little rest.

Knowing what I learned on the trip, would I have climbed Kilimanjaro in the first place? No.

But I'm glad I did it. It was the ultimate goal. I pushed myself way beyond what I thought I could do, and I made it most of the way to the top.

And that's pretty much how I feel about my life.